GREAT AVIATION STORIES

VOL. 1

MICHAEL BARRY

Saturn Books, Fermoy, Co. Cork

By the same author:

No Flowers By Request
Poems For Your Pleasure
An Affair of Honour
The Romance of Sarah Curran
The Story of Cork Airport
International Aviation Quiz Book Vol. 1
By Pen and Pulpit
The Mystery of Robert Emmet's Grave
International Aviation Quiz Book Vol II

© Copyright Michael Barry 1993

ISBN 0 9515387 4 8

Front Cover Photograph: - P.10 Vega Gull.
© Austin J. Brown, Aviation Picture Library, London.

Printed by Litho Press Co., Midleton, Co. Cork.

ACKNOWLEDGEMENTS

The author gratefully acknowledges the financial assistance given by AER RIANTA towards the cost of research, photographic material etc. in connection with this book and a special word of thanks in this regard to Mr. Barry Roche, General Manager, Cork Airport.

Sincere thanks is also expressed to the following:- The United States Navy (Navy Historical Centre); The Henry Ford Museum, Dearborn, Michigan; The Imperial War Museum, London; The National Air & Space Museum, Smithsonian Institute, Washington; The Royal Air Force Museum, Hendon; Short Bros. Plc, Belfast; The Zeppelin Museum, Friedrichshafen, Germany; The Reader's Digest Association, London; Federal Aviation Administration, Washington; Lufthansa Airlines, Germany; Aer Lingus; The American Embassy, Dublin; The Cork Examiner; The German Embassy, Dublin; Stephen Coughlan, Journalist, Cork; Ger. O'Regan, War Plane Research Group, Ballinlough, Cork; Lieut Col. K.T. Kelly, Irish Army; John Cussen, Solicitor, Newcastle West; Eugene Pratt, Aer Rianta, Shannon Airport.

The author owes a deep debt of gratitude to Gabriel Desmond for his sound advice and observations on reading the MS.

Grateful thanks is also due to the New York Public Library, The National Library of Ireland and the Cork City and County Libraries. A very special thanks to Tim Cadogan, Reference Section, Cork Co. Library, to Miss Mary Neville of the Fermoy Co. Library and to Tony Storan of Limerick Co. Library for their kind assistance and co-operation.

REFERENCES

The main references are as follows:- The World's Greatest Air Mysteries - Michael Hardwick (London 1970); Great Mysteries of the Air - Ralph Barker (Chattis & Windus 1966); Playboy of the Air - James Mollison (Michael Joseph 1937); Wings of Mystery - Riddles of Aviation History (Titler, Dale, New York 1966); Women and Flying - Lady Heath and Stella Wolfe Murray (John Long 1929); Women of the Air - Judy Lomax (John Murray 1986); Ace With One Eye- Oughton & Smyth (Muller 1963); Personal Diary of "Mick" Mannock; Major Edward "Micky" Mannock VC. DSO. MC - Capt. J. Morris (RAF Quarterly Jan. 1932); "Mick" Mannock -The Forgotten Ace - Thos. J. Mullen, J.R. K.M. (Irish Sword); The Stars at Noon - Jacqueline Cochran (Brown, Little 1954); Ireland and World Aviation - L.M. Skinner & T. Cranitch (Direct Publications (Ireland) Ltd. 1988); The Airship - A History - Basil Collier; Strange Encounters, Mysteries of the Air- David Beaty (Metheun 1982); Winged Legend - The Story of Amelia Earhart (John Burke, London 1971); The Search for Amelia Earhart - Fred Goerner (New York 1966); Last Flight - Amelia Earhart - Edited by G. Putnam - (New York 1937); Flight into History - Elsbeth E. Freudenthal (University of Oklahoma Press 1949); Famous Flights - J.F. Turner (Barker 1978); The Hindenburg - Michael M. Mooney (Hart-Davis 1972); Who Destroyed the Hindenburg - A.A. Hoehling (Brown, Little & Co. Boston 1962); Yeager - Gen. Chuck Yeager and Arnold Schwartzman; A Time to Fly - Sir Alan Cobham (Edited by Christopher Derrick) - Shepheard-Walwin 1978; The Aeronautical Journals (London); Century Magazine; Slip Stream; Harpers Magazine Nov. 1937; Reader's Digest, Nov. 1937; Aviation Digest, September 1987; The Irish Sword; An Cosantoir; Irish, British and American Newspapers; Newcastle West Observer - Robert J. Cussen; The First Flight Across the Atlantic, May 1919 - Commander Ted Wilbur (Smithsonian Institution- 1969)

ILLUSTRATIONS

Col. Fitzmaurice with *Bremen* crew.	13
The *Bremen* crew being welcomed in Berlin.	17
The *Bremen* in the Henry Ford Museum, Michigan.	19
Douglas Corrigan's Curtiss Robin plane at Baldonnel, 1938.	23
Douglas Corrigan in the American Legation, Dublin 1938.	25
Douglas Corrigan with American Minister in Dublin 1938.	27
The *R101*.	35
The wreckage of the *R101*.	41
Amelia Earhart with Mr. McCallion in Derry, 21st May 1932.	49
Amelia Earhart with her aeroplane on that same occasion.	51
The Lockheed Vega - Amelia Earhart's plane.	53
Amelia Earhart and her Lockheed Electra.	57
Lady Mary Heath.	65
The Wright Brothers.	73
The first successful heavier-than-air flight.	77
The 'Wright Flyer' 1909.	79
Page from Short Bros. Plc Order Book in 1909.	81
Lieut. Commander A.C. Read - US Naval Aviator No. 24	87
The crew of the NC-4.	89
The NC-4 at Lisbon in May 1919.	93
"Mick" Mannock.	101
Jacqueline Cochran.	113
Alan Cobham with his Air Circus in Cork 1933.	121
The *Hindenburg* in flames.	135
Jim Mollison in New York 1932.	145
Mr. Crosbie's ascent in a balloon from Leinster Lawn, Dublin.	153
James Sadler, first English aeronaut.	155

INTRODUCTION

From the very early days of ballooning to the space age of today, aviators and their flying craft have been a considerable source of interest, and in this book we look at the lives and the achievements of some of those famous flyers, especially in the early decades of this century when many flew literally "by the seat of their pants".

In *Great Aviation Stories* we meet some of those aviators who are now part of aviation history such as the two bicycle mechanics - the Wright Brothers who, having successfully made the first flight in a heavier-than-air machine in December 1903, set aviation on the road to the advanced state it is in today,

Also included are stories about great flyers whose feats were often eclipsed by those of their much publicised contemporaries. One such story concerns American Lieut. Commander Read and his crew who, in a flying boat, actually made the first crossing of the Atlantic in a heavier-than- air machine in 1919, stopping en route at the Azores to refuel. This outstanding achievement merited far greater recognition than it ever got but was masked by the huge publicity which surrounded the first non-stop crossing of the Atlantic some few weeks later by Alcock & Brown.

Included too are the behind the scene stories such as that of the Airship *R101* which, for political expediency, set out on a flight from England to India in October 1930 without ever meeting the full terms of a Certificate of Air Worthiness and crashed in Beauvais, France with great loss of life. We look at the great Amelia Earhart's famous flights, especially her last. Did she crash and disappear without trace or did she come down on or near one of the Japanese-held islands in the Pacific where she and her co-pilot were picked up and held by the Japanese and later died? There is good reason to believe that this may have been the case. It is examined here in some detail together with the involvement that American Intelligence may have had in the flight. The tragedy of the *Hindenburg* in 1937 is also dealt with and the part that sabotage against the German Third Reich may have contributed to the airship going up in flames as it prepared to land.

There are also the stories of famous Irish aviators such as that of Col. James Fitzmaurice who, with two Germans, made the first east-west crossing of the Atlantic in the *Bremen.* Fitzmaurice, an outstanding airman, never got the honour he deserved in Ireland, yet his two German colleagues are legends in their own country. Then there is that gallant Irishman Major Edward "Mick" Mannock, V.C. who, without sight in his left eye, bluffed his way into getting a transfer from the

Royal Medical Corps to the Royal Flying Corps and became the top flying ace with the British forces in the First World War. The story of Lady Mary Heath, the former Sophie Pierce from Newcastle West, Co. Limerick is told. She made the first solo flight by a woman from South Africa to England in 1928 and before her flying career blossomed, was a fine athlete, creating the world high jump record for women in 1923.

One can read of Sir Alan Cobham, the air circus man and of the major contribution he made to the development of commercial aviation. Also included is the story of Jacqueline Cochran who, from extremely humble beginnings and with little or no formal education, became not alone a highly successful business woman but also an outstanding aviator who went on to become the first woman in the world to fly faster than the speed of sound.

We look at the great flying achievements of Jim Mollison who always had an eye for a pretty girl. He was the husband of Amy Johnson for some years and made the first east-west solo flight across the Atlantic from Portmarnock Strand in Dublin. Did Douglas "Wrong Way" Corrigan make a compass reading error and cross the Atlantic to Dublin from New York in July 1938 instead of flying west to Los Angeles as per his filed Flight Plan or did he deliberately make the flight eastward.? His story is well worth recalling.

We look at the great balloonists such as the Montgolfier brothers who made the first hot-air balloon flight in April 1783, thus giving man that great freedom to take to the skies. Ireland too was involved in the early years of this type of aviation, the great Richard Crosbie being very much to the forefront.

There are thirteen stories in all, each with its own distinctive background, together with many extremely interesting photographs. I hope you enjoy them.

Michael Barry.

CONTENTS

	Page
Introductory Pages	1 - 8
Fitzmaurice - The Forgotten Irish Aviator	9
"Wrong-Way" Corrigan	21
The Flight That Should Never Have Been Made	30
Amelia Earhart - Her Mysterious Disappearance	44
Lady Mary Heath	64
The Bicycle Mechanics Who Made Aviation History	68
The First Atlantic Crossing	84
"Mick" Mannock - The Irish Flying Ace	95
From Slum To Sky	106
Cobham - The Air Circus Man	116
The Hindenburg	128
Romance in the Air	140
Balloons and Balloonists	150
Index	157

FITZMAURICE - Ireland's Forgotten Aviator

When Commandant James C. Fitzmaurice as co-pilot of the *Bremen* with its German pilot Captain Herman Köhl and third man Gunter Von Hünefeld landed safely at Greenly Island off Labrador on Friday 13th April 1928 they made aviation history for having made the first successful East-West crossing of the Atlantic.

Fitzmaurice was born in Portlaoise in 1898 and received his early education at the local Christian Brothers School. He was later sent to Rockwell College. After some time there he left and took a job for a short while as an assistant in Hearne's drapery establishment in Waterford. At the outbreak of war in 1914 he joined the Cadet Company of the 7th Leinster Regiment without his parents' permission. He was then only 16 and when his father discovered what he had done he had him removed. However, the following year he joined up again, this time with the 17th Lancers and went with them to France. He became a sergeant at 18 and while serving there was twice wounded. He was later recommended for a commission and posted back to England for training in a Cadet College.

When commissioned Fitzmaurice was posted to the 8th Battalion of the Liverpool Irish Regiment. By this time he had developed a keen interest in aviation and requested a transfer to the Royal Flying Corps. He served with them until 1919. While training to become a pilot he had met and later married an NCO from the Women's Royal Air Force. In May 1919 he was chosen to make the first ever night mail flight from Folkstone to Boulogne. He was later selected as a pilot on a proposed flight from Cairo to Capetown but this was later abandoned.

When he was demobbed he entered the insurance business. He was very unsettled in that job and joined the RAF on a short service commission in June 1921, coming out again in 1922. He then came back to Ireland with his wife and baby daughter. The Irish Air Service was then being formed and he applied for a post as pilot. He was given an interview and following a flight display was accepted and commissioned. He began training pilots for the new Service. In September 1922 he was given command of Fermoy Aerodrome in Co. Cork. In 1923 he was promoted Captain and became Commanding Officer of the Irish Air Corps in 1926.

An aeroplane crossing of the Atlantic, the dream and inspiration of countless aviators in the early years of flying and one which sent many to their deaths was first successfully made in 1919 in a West-East direction. Alcock and Brown who made the first non-stop flight in that year achieved enormous publicity when they landed at Clifden in Co. Galway where there is now a monument to their memory. However, eight years went by before further serious efforts for an Atlantic crossing were made. Lindbergh in May 1927 made the first successful solo West-East Atlantic crossing and in June of that same year another successful crossing was made by Clarence Chamberlain and Charles Levine. In fact ten attempts had been made that year with only two successes. The eight failures had resulted in the loss of nine lives and three aircraft.

No one had yet made a successful East-West Atlantic flight. This was looked upon as the more difficult crossing due to the prevailing westerly winds opposing the direction of flight. There were those from the mid 1920s onwards willing to take the risk even though it was looked upon by many in aviation as suicidal. Fitzmaurice was one who was prepared to do just that.

While ten attempts had been made in 1927 in an West-East direction. seven unsuccessful East-West trans-Atlantic attempts were also made and five lives were lost. One of the attempts was that on Sunday 14th September when two German planes, the *Bremen* and the *Europa* took off from Dessau Aerodrome of the Junkers factory to fly to North America. After a severe battle with the weather for several hours, the pilots of both aircraft were forced to abandon their attempts and return. The pilots of the *Bremen* were Captain Köhl and Herr Fritz Loose and also on board was Herr Von Hünefeld as passenger. Flying the *Europa* were pilots Edzard and Risztics who had an American journalist Knickerbocker as passenger.

One month later on 16th September Commandant James Fitzmaurice teamed up with Captain McIntosh a famous English pilot from Imperial Airways, and affectionately known as "All Weather Mac" in an attempt to make the East-West crossing. After four hours flying in very bad weather they were forced to return and managed to make a successful landing on Beale Strand in Ballybunion, Co. Kerry

In the following year 1928 efforts to make a successful East-West crossing of the Atlantic were intensified despite the very strong warnings of the dangers involved. The warnings to a great extent went unheeded. In March, Captain Walter Hinchliffe and the Hon. Elsie Mackay daughter of Lord Inchcape who was Chairman of the P&O Shipping Line set out. The couple were lost without trace, the last sighting of their plane was over Ballydehob in South West Cork, flying on a westerly course.

The unsuccessful German attempt at an East-West crossing in which

Köhl and Hünefeld were involved only intensified their intention of trying again and it is with this second attempt that Commandant Fitzmaurice became involved. The German press by and large looked upon them as mad and conveyed the message that enough lives had already been lost and that Köhl and Hünefeld were lucky to have survived their first attempt.

Any new serious attempt at an East-West Atlantic crossing would require enormous preparation and most of all finance. Much had been learned about the *Bremen's* performance from the failed attempt in August 1927. It had behaved extremely well under atrocious weather conditions. One way to reduce the risk factor on a new attempt would be to begin the flight from Ireland.

The man behind the entire project was Gunther Von Hünefeld aged 36 from East Prussia. He had a brilliant mind but never enjoyed great health. He had been interested in aviation from an early age but could never become a pilot as he was blind in one eye. He won the Iron Cross II in the First World War and in 1923 became Press Chief of North German Lloyd where he conceived the idea of an East-West attempt to cross the Atlantic. He greatly admired Köhl whose ability as a pilot to lead such an attempt was never in doubt.

Captain Herman Köhl was aged 40 and a native of Bavaria. He had been a pilot in the First World War and had afterwards become Chief of Lufthansa Night and Bad Weather Flying Unit.

Throughout the winter of 1927 work went ahead on the *Bremen*. The German Aircraft Builder Professor Junkers placed his workshop again at the disposal of Köhl and Hünefeld to facilitate the engineering work necessary to achieve success. German Lufthansa released Köhl for the venture that lay ahead. One of the main obstacles to the success of the operation was finance. On this occasion the Junkers factory in Dessau could not afford to take on new financial burdens due to the failure of the initial attempt. Accordingly it was left to Hünefeld to raise the capital and handle the administration involved while Köhl in close co-operation with the Junkers factory would prepare the *Bremen*. The capital was raised and work went ahead.

As already mentioned, the intention was to begin the crossing attempt from Ireland. Von Hünefeld made a private approach to the Irish Government seeking permission to use some aerodrome location there as a take off point and a close friend of his named Herr Klose who was a Director of North German Lloyd made a formal approach. Permission was readily granted and all possible assistance was offered. The three men Köhl, Hünefeld and Klose travelled to Ireland at the end of February 1928 to study likely aerodromes.

The Germans had considered the airstrip at Oranmore in Co. Galway. However, it was then in a very disused state and would cost a considerable amount to make it suitable for a heavily laden plane to take off. Baldonnel was suggested as being an ideal location and this was accepted following an inspection. It was there that the Germans had met Commandant James Fitzmaurice, Officer Commanding the Air Corps. Writing later about this Hünefeld recorded:-

"The reception given us by the Irish Air Corps was overwhelming. Köhl verified the suitability of the runway immediately thus setting aside the last obstacle. But we had to go cautiously and methodically. The machine was completely overhauled and every detail attended to. I even managed to charter a plane (of similar type to the *Bremen*) for Köhl in which he was able to carry out extremely important tests in Templehof-Berlin together with his one-time pilot Unter Offizer Spindler."

Spindler had also been chosen by Köhl to be his co-pilot on the proposed Atlantic crossing attempt but as we shall see later his place was given to Fitzmaurice.

Strict secrecy was being observed by the German crew regarding the *Bremen's* new attempt. The German Government had advised all airport managers to prevent any aircraft from taking off on what they considered to be a suicidal mission across the Atlantic. Mindful of this Köhl and Hünefeld worked out a strategy. On 26th March 1928 they filed an internal flight plan for the *Bremen* from Dessau to Berlin. They did not of course land in Berlin but headed for Ireland.

The Air Corps authorities were expecting the flight and preparations were made accordingly. Around 4.20 pm the Junkers monoplane *Bremen* appeared in the distance. Commandant James Fitzmaurice went up in a Moth aircraft to indicate the landing area and he then landed again. The *Bremen* circled the aerodrome for some time, its markings BREMEN D-1167 on its silver grey body being easily readable. Fitzmaurice went up a second time and made another landing followed by the *Bremen* at 4.30 pm. The aircraft was immediately put in a hangar by aerodrome mechanics.

When it was learned in Germany that the *Bremen* had landed in Ireland and intended to attempt another East-West Atlantic crossing from there, the reaction was at first very strong and extremely critical. The German press was loud in its condemnation of what the aviators had done and expressed little hope of the success of their intended

Baron Von Hünefeld (left), Major James C. Fitzmaurice and Hermann Köhl pictured with the Bremen.
Courtesy Lufthansa

flight.

The arrival of the *Bremen* at Baldonnel on 26th March 1928 was an event which, although unknown to Fitzmaurice at the time, would write him into the pages of aviation history. As Commanding Officer of the Air Corps he had made the German crew very welcome. Recording this event later Hünefeld wrote

"The welcome given us by the Irish military and civil authorities was most hearty. We were made guests of the Irish Air Corps and soon it was settled between Köhl and me to invite Commandant Fitzmaurice to accompany us as second pilot on our flight to North America. And so our 'German-Irish Crew' as we came to call it was formed and none of us ever regretted the pact which proved itself so trustworthy in the course of extreme danger."

The Germans had brought over with them two mechanics to ensure that everything would be in order when the aeroplane took off on such a daring flight. They were assisted in every way by four mechanics from the Irish Air Corps, Johnny Maher, Sid Peacock, Leo Canavan and George Barton.

It was necessary to have the longest run possible for the *Bremen's* take off as the aircraft would be carrying maximum fuel at that time. To achieve this, alterations had to be made. The take off end was a concrete area between two hangers but close to this was a small dip in the ground which Fitzmaurice arranged to have filled in. To get the maximum length at the other end he had a wall bordering the aerodrome knocked down. That field next to it was later to be called the "Bremen" field.

That Fitzmaurice had been asked to join the Germans in their effort to be the first to make a successful East-West Atlantic flight was an enormous vote of confidence in the Irish pilot. It was also a wonderful opportunity but a flight that would be fraught with danger. He had a wife and child to consider. He applied for permission from his superiors to undertake the flight which was granted. The fears he had for his family should he not survive were allayed by a government assurance that if this were to be the case his gratuity and pension would pass to his wife.

Once the personal matters were safely out of the way and Fitzmaurice was free to undertake the flight, the news broke and attracted wide attention. By this time Lieut. Spindler, who was originally to be the co-pilot, had returned to Germany. Some weeks had elapsed since the *Bremen* had first landed at Baldonnel. A number of trial runs were undertaken to determine the precise point the aeroplane should be on the runway before lift off.

Following a favourable North Atlantic weather report, dawn on Thursday 12th April was the time and date set for the *Bremen's* departure. The plane was expected to lift off around 5.30 am. From early hours large crowds began to flock to Baldonnel, all eager with the anticipation of witnessing Irishman Fitzmaurice and his two German colleagues endeavouring to make aviation history.

Everything was in readiness for the flight around 4 am, some fifty soldiers having wheeled the machine out of its hanger. The runway had been flagged to give the pilots direction for take off. The weather was fine and the sky cloudless. The aerodrome was well lit and there was a sense of great activity. Provisions for the flight had been put on board which included seven thermos flasks of tea and beef tea, a supply of chicken and beef sandwiches, some chocolate and six peeled oranges. A strong military presence prevented crowds from getting too near the *Bremen*, a ghost-like machine in the breaking dawn. It was powered by a Junkers L.5 engine of 310 horse-power. It carried 600 gallons of fuel and fully loaded the plane weighed 5 tons. Also on board were two copies of the *Irish Times*, the first European paper to cross the Atlantic.

Among those who arrived at Baldonnel to wish the plane and its crew God speed were President and Mrs. Cosgrave, The Minister for Defence Mr. Desmond Fitzgerald, the Army Chief of Staff Major General Hogan and the German Consul General.

Around 5.15 am the plane's engine was started. The Irish Tricolour and the German Imperial Flag of black, white and red flew from the *Bremen* while inside Captain Köhl, Commandant Fitzmaurice and Baron Von Hünefeld made their last minute adjustments. At 5.38 am the plane took off bound for New York. Everything went well as the plane gathered speed for lift off. Just as it was approaching the point on the runway where lift off would occur a sheep suddenly broke loose from a distant herd which had been rounded up in a corner of the aerodrome. The animal was heading directly on to the path of the approaching *Bremen*. At this time Captain Köhl was looking ahead and also to his left and did not see the sheep. Commandant Fitzmaurice who was looking to his right saw the animal about to run across the plane's path and immediately grabbed the joy stick and pulled it back. The plane began to lift and missed the sheep. Although lurching a little, the *Bremen* remained in the air and the aviators were on their way. What could have been a disastrous collision had been averted by Fitzmaurice.

As the plane flew out over Galway heading westward the weather was calm and all looked well. In fact for the first 15 hours or so the crew experienced little difficulty. From time to time they threw out smoke bombs to calculate wind speed and direction. They had a considerable amount of daylight as they were flying to the west. Their destination was to be New York, flying a Great Circle track. They had estimated that they would reach Newfoundland around 4 am the following day arriving in New York about 8 am.

As they headed towards the American Continent the weather began to deteriorate and strong head-winds made their presence felt. This reduced their speed. They were now flying in complete darkness and in icy gale force winds. To add to the crew's problems the plane's oil tank developed a leak. Fitzmaurice managed to get out of his seat and crawl back. He eventually found it and carried out a repair. This steadied the oil gauge again. Hünefeld later recalled:-

"By this time the storm rose to such a fury that the plane appeared to make no headway. Cloud and fog patches raced past. The furious sea sought to tear itself asunder. Memories of the first stormy flight of the *Bremen* bore in on us. Now she simply was not flying anymore but dancing. But she held together and battled forward. ..."

By now the plane was being driven northwards off its normal course and in the darkness it was impossible to get any idea as to where they were. It was only when daylight came and they were able to see the snowy wastelands of Labrador beneath them that they knew they were well off track. They altered course and hoped that they could identify some land mark. Fuel too was beginning to run low. Suddenly Fitzmaurice shouted that he could see a boat. It was actually a house. As they came lower they flew over a lighthouse and shortly afterwards Köhl set the aeroplane down on a frozen surface. Under the circumstances the landing was excellent but as the aeroplane came to a halt it tilted forward on its nose and damaged the propeller.

At least the crew were safe and had made aviation history by being the first to make a successful East-West crossing of the Atlantic. They had landed the *Bremen* on Greenly Island off Labrador in an area where there was a Canadian fishing station in the Straits of Belle Isle, north west of Newfoundland. Greenly Island was 400 miles (644 km) from any point over which the aeroplane was calculated to pass on its way to New York.

News of the successful East-West crossing of the North Atlantic was greeted with jubilation throughout the world. Heads of State from many nations extended their congratulations to Germany and Ireland

Köhl, Fitzmaurice and Hünefeld being enthusiastically greeted in Berlin following their return from America
Courtesy Lufthansa

on the success and gallantry of their airmen and overnight the crew of the *Bremen* had become world heroes.

In Ireland, Fitzmaurice's part in the great achievement was greeted with elation and pride. President Cosgrave cabled his congratulations to him as did the Minister for Defence, Mr. Desmond Fitzgerald who advised him that he had been promoted to the rank of Major.

Later Major Fitzmaurice left Greenly Island for some time in a relief plane to get some spare parts for the *Bremen*. When he returned, he learned with great sadness that Floyd Bennett the Polar Airman who was actually co-pilot of the first plane to reach them after their landing on the ice, had contracted pneumonia in the severe Arctic conditions of Greenly Island, from which he died. To show their appreciation of what Bennett had done, Fitzmaurice and his German colleagues set out for Washington DC in the *Bremen* which was now air worthy again to attend the funeral. The weather proved treacherous and for safety sake they landed in New York instead and proceeded by car to Arlington cemetery where Floyd Bennett was being buried.

The reception the airmen received in New York was fantastic and this was repeated in many other American cities. Mrs. Fitzmaurice and her young daughter together with Mrs. Köhl had travelled to America by liner and received the adulation of the enormous crowds everywhere they went. It was a continuous round of engagements for all concerned. Mrs. Fitzmaurice had brought with her from Dublin her husband's dress uniform. Before leaving America Major Fitzmaurice presented his sword to the New York Museum and most important of all, Baron von Hünefeld presented the *Bremen* to the American people where to day it is proudly on display in the Henry Ford Museum, Dearborn, Michigan.

On Saturday 16th June the German and Irish party arrived back from America by liner at Plymouth and from there went on to Germany where they received the accolades of a proud nation. On Monday 2nd July the three airmen flew from Hamburg to Croydon in the *Europa*, a sister ship of the *Bremen*. In Croydon they were entertained to lunch by the Aero Club. That night the Union of Four Provinces Club gave a banquet in their honour and on the following evening the three airmen landed at Baldonnel, the aerodrome from where they had set out on their epic flight.

Among the honours and gifts bestowed on them was the Freedom of the City of Dublin. President Cosgrave held a reception in their honour at the Metropole Hotel in Dublin while at a dinner in their honour in McKee Barracks, the Chief of Staff Major General Hogan presented each of the airmen with an Irish sword. They were also

The Bremen in the Ford Museum, Dearborn, Michigan.

Ford Museum, Dearborn.

honoured guests at a Garden Party in the American Embassy. On 24th July "Fitz" as Major Fitzmaurice was affectionately called was promoted to the rank of Colonel.

Ireland then was proud of Colonel Fitzmaurice who had brought his country honour and glory. Anyone in Ireland at that time could sense that pride and would have felt assured that the Irish airman's fame would live on. Sadly this was not so. Fitzmaurice was a keen aviator and saw the great potential commercial aviation could have for Ireland. When his efforts to interest his government in air services between Ireland and Britain failed, he got disillusioned and resigned from the Air Corps.

He spent some years in America and returned from there in 1939 and went to live in England where he ran a service men's club during the war years. He came back to Ireland in 1951 and died in Dublin on 26th September 1965.

Baron von Hünefeld died the year following their epic flight and in 1937 Capt. Herman Köhl passed away. Today they are proudly remembered and honoured in their own country for their outstanding achievement. Colonel James Fitzmaurice however is Ireland's forgotten aviation hero.

"WRONG-WAY" CORRIGAN

One of the great unresolved air-mysteries is that of the flight by American airman Douglas Corrigan in July 1938 from New York's Floyd Bennett Airfield - destination west to Los Angeles but ultimately landing at Baldonnel Aerodrome in Dublin, having crossed the Atlantic in an easterly direction.

Douglas Corrigan, then aged 31, was a keen aviator. He was also an aeroplane mechanic who, eleven years earlier while employed with the Ryan Aircraft Factory in San Diego, California, had worked on the preparation and servicing of Lindbergh's plane *The Spirit of St. Louis* in which he flew the Atlantic in 1927. It was a flight that Corrigan hoped that he too would achieve one day but such flights took a great amount of planning and of course required much financial backing. Also, aviation authorities were reluctant to sanction such non stop flights especially over water due to the risks involved.

Corrigan was the owner in 1938 of a Curtiss Robin monoplane. It was a 165 hp. machine and although he kept it in the peak of condition due to his sound mechanical knowledge, it was nevertheless nine years old and single engined. He had extra fuel tanks fitted forward of the cockpit which really made it impossible for him to see directly ahead. He was obliged to turn broadside to get a forward view. Lindbergh faced a similar problem for his flight across the Atlantic but he had a periscope fitted which overcame it. The added fuel tanks gave Corrigan about 36 hours average flying time. The plane had little in the line of instrumentation other than a compass, a bank indicator and an inclinator which enabled him measure the steepness of climbs and descents.

On Sunday 10th July he flew from Los Angeles non stop to Floyd Bennett Airfield in New York, a distance of 2,700 miles (4,345 km) and which took him 27 hours 50 minutes. This was a great achievement at that time and the flight attracted some publicity. Had he not made it and was forced to come down en route, the aviation authorities would in all probability have revoked his licence for the return flight. Also, the plane did not exude any confidence in them.

Corrigan stayed a week in New York servicing and preparing his plane for the return journey to Los Angeles. Kenneth Behr who was the manager of Floyd Bennett Airfield didn't rate Corrigan's plane very highly. The latter had advised the manager of his intention to leave New York for Los Angeles on the return leg of his flight at midnight on

Saturday 16th July. Behr however refused him permission to leave at that time but gave him clearance to take off at dawn the following morning. The weather forecast for the westerly flight was satisfactory.

And so at first light on Sunday 17th July, Douglas Corrigan took to the air having filed a flight plan for Los Angeles. With full tanks of fuel, the nine year old single-engine plane was sluggish in attaining altitude due to the fact that it was half a ton overweight. Visibility was not great on take-off due to haze and patchy fog. The runway that Corrigan chose for take off would put him on a course over Long Island but when sufficient altitude was obtained he would then swing round to pick up a westerly course for Los Angeles.

His story later was to be that, at a height of 500 ft., he began a slow turn to starboard but found that his compass appeared to have developed a fault and wasn't behaving as it should be. Luckily he had brought a spare compass with him and which he had set up before take off. He followed this compass readings until he was satisfied he was on a westerly course for Los Angeles. Actually he had misread the compass and was flying on a reciprocal bearing which would eventually take him out over the Atlantic. Because of the banks of cloud, Corrigan was unable to see any distinguishable landmark and he continued his climb. A few momentary breaks in the cloud gave opportunity to some people on the ground to catch a glimpse of the small plane that for some reason wasn't heading west but was on an easterly course.

To the aviation officials that meant one thing - Corrigan had duped them into thinking that he was making a return flight to Los Angeles as per his flight plan but instead he was attempting a solo flight across the Atlantic. It was a flight in a nine year old monoplane that had 45 horse power less than that of Lindbergh's plane. Corrigan's plane was of a type primarily built for training purposes. The authorities felt that he had no financial backing whatever. In fact, he was not at all well off and during the week he spent at Floyd Bennett Airfield following his arrival from Los Angeles, he had slept in a hangar beside his plane to save on hotel expenses.

Would Corrigan be so daring and foolhardy to undertake such a journey over water? Officialdom was more than suspicious. He could make it with the fuel he had on board with the prevailing westerly wind behind him. He had not taken too kindly to the Bureau of Air Commerce's persistent refusal in allowing him to attempt an Atlantic crossing and had let it be known to some of his close friends. He had casually remarked to them on one occasion on a way to beat the system. "Why not land at Floyd Bennett Airfield late one evening when senior

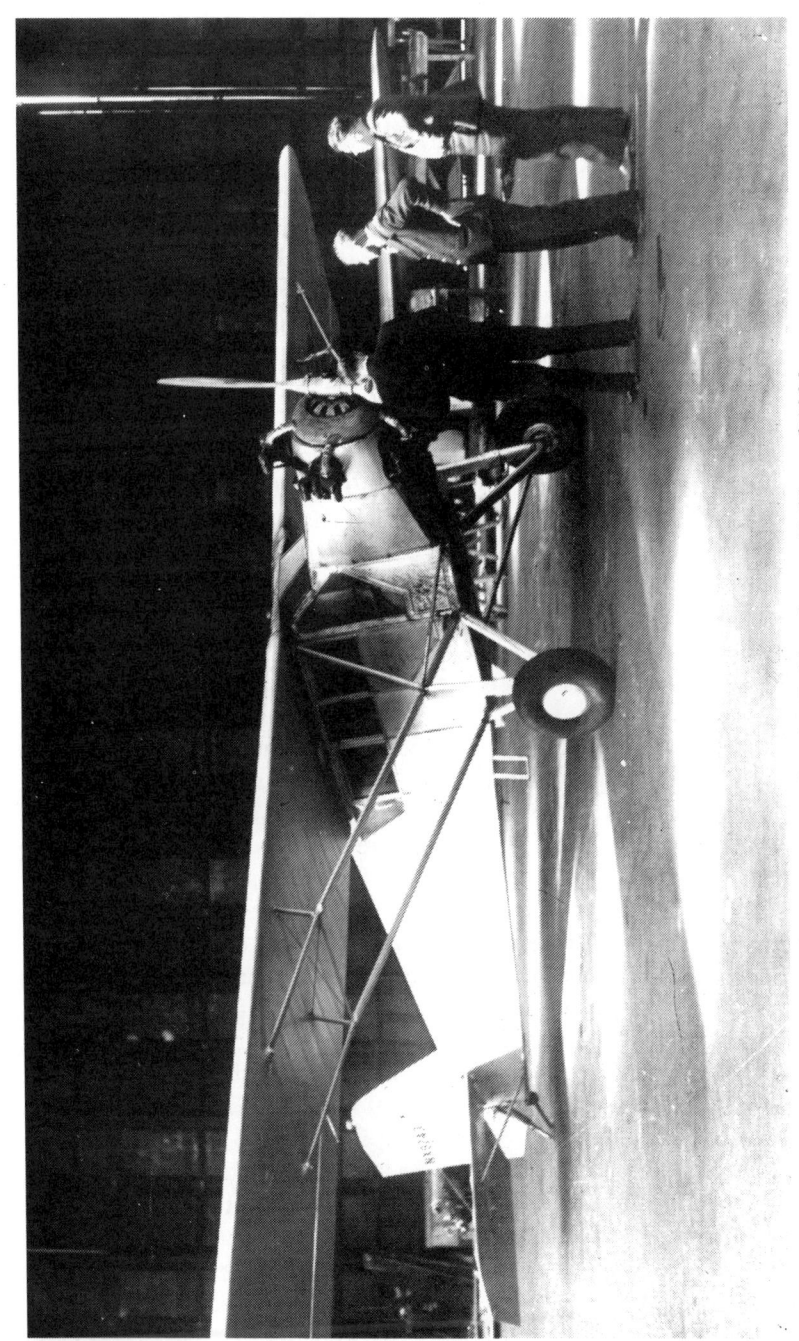

Douglas Corrigan's Curtiss Robin plane at Baldonnel 1938

Courtesy Irish Times

officials have gone home, fill up with gasoline and just go?"

However, Douglas Corrigan was not a reckless airman and an Atlantic flight was fraught with danger. His plane did not have a radio and he had not been given nor did he seek Atlantic weather reports. The only map he carried referred to his route from New York to Los Angeles across America via Memphis and El Paso. On the face of it therefore, it would be suicidal to attempt a flight eastward. His closest friends knew nothing whatever of any intention of his to make an attempt to cross the Atlantic. He didn't carry a passport, surely an indication that he didn't intend to leave America. Furthermore, when he had landed at Floyd Bennett Airfield he discovered that one of his fuel tanks had developed a leak. He decided to take a chance and have it repaired on his return to Los Angeles as its removal for welding would delay his return by some days. At least, should he run into difficulties on the way back, he was hopeful of finding a place to land. There would be no such place over water. Therefore taking everything into consideration it would have been grossly irresponsible of him to deliberately take an easterly track. The Bureau of Air Commerce were not convinced. There had been no sightings of the plane if it had eventually swung back on a westerly course and the conclusion by them was that Corrigan was on an attempted Atlantic crossing without permission.

The poor and patchy ground visibility militated against him discovering his navigational error. When he did get a brief glimpse below him he was actually over Boston which he erroneously mistook for Baltimore. Slowly but surely the plane began to gain height as he burned off his fuel and he was happy with his progress. He was over the northern tip of Newfoundland after about ten hours flying but through a momentary break in the cloud he saw unidentifiable land.

It was after that he got the first jolt that would require all his ingenuity. The leaky fuel tank appeared to be getting worse and petrol began to seep into the cockpit. It was a scary situation but to Corrigan, in the event of a serious emergency, he was confident that he would get the plane safely down on land. One real fear he had was that of fire. He was sitting on a time bomb in a cockpit made of wood. To alleviate his predicament, he bored a hole with a screwdriver through the cockpit floor on the opposite side from the exhaust pipe and the petrol drained off.

From time to time he estimated his position but confirmation due to cloud was out of the question. He was now flying at 8,000 feet and the cloud began to thicken as he flew on during the night. It had also begun

Douglas Corrigan in the American Legation, Dublin 1938
Courtesy Irish Times

to rain. This later turned to sleet and the plane's wings began to ice up. He was not fitted with any de-icing equipment so he dropped the plane down through the clouds to 3,500 feet to overcome the problem, expecting to see land below. To his utter amazement he saw the sea below him. Momentarily he thought that perhaps he had made such good time that he had overflown land and was over the Pacific Ocean but on calculating his flying time that wasn't possible. He re-checked his compass only to discover that he was on an easterly track and it wasn't the Pacific but the Atlantic that he was over. Terror swept over him. Douglas Corrigan, the experienced airman had made the mistake not unknown to beginner pilots of reading the reciprocal bearing and not the direct one.

As of now he didn't know where he was other than he was somewhere over the Atlantic and that hopefully the Great Circle track he was flying would put him over Ireland. He did some elementary calculations and flew on. His first sighting was that of a small trawler, an indication that he couldn't be too far from a coast line. After flying for 27 hours he eventually picked up the west coast of Northern Ireland. He continued to fly over land until he crossed to the east coast where below him he saw a large city. He was unable to locate any airfield there and so he altered course and flew down the coast hoping that he would eventually pick up Dublin. He had studied this area previously in connection with his hoped for Atlantic crossing and knew of Baldonnel Aerodrome. As he flew down the coast line a military plane came close to identify him and seeing that he posed no threat it flew away. Eventually he flew over Dublin and landed at Baldonnel.

Word had been sent to the Irish Authorities via the American Legation that Corrigan had flown east instead of west following his departure from New York so his arrival did not come as a complete surprise, although he had no permit to land. On satisfying customs and immigration officials he later left for the American Legation in the Phoenix Park.

News of his arrival spread rapidly. When interviewed he said " I never intended to fly across the Atlantic. When I left Floyd Bennett Airfield yesterday, I was going to fly to Los Angeles but when I got above the clouds something went wrong with my compass. He stuck consistently to his story in the many interviews he was later to give both in Ireland and in America. Mr. Kenneth Behr, manager of Floyd Bennett Airfield on learning of Corrigan's arrival in Ireland said "Considering his equipment and everything, I guess it is the greatest flight we have ever had. That was a terrible ship he was on."

Douglas Corrigan with Mr. John Cudahy, American Minister to Ireland 1938
Courtesy Irish Times

Corrigan's relatives and friends on hearing of his great achievement were also not convinced that he had made an error. His uncle, the Rev. Frazer Langford when interviewed at Los Angeles remarked that "Corrigan was most reticent just like Lindbergh. He always kept his plans to himself until he had carried them out. He is a great admirer of Lindbergh you know and that is a great plane, even though it is nine years old. He kept it ticking like a watch. I am most proud of the boy. I knew he had skill and courage."

Corrigan stayed in Dublin for several days as guest of the American Legation and while there met the President of Ireland Mr. Sean T. O'Kelly and An Taoiseach Mr. de Valera. As a result of his flight the American Bureau of Commerce suspended his Experimental Aircraft Certificate in respect of the plane. Colonel J. Monroe Johnson, Acting Secretary of Commerce, had sent a letter to Mr. Cordell Hull, U.S. Secretary of State, asking that the suspension order be delivered to Corrigan and to Mr. John Cudahy, American Minister in Dublin. Colonel Johnson explained that Corrigan, by flying the Atlantic had violated the terms of the experimental certificate which had only authorised a flight from Los Angeles to New York and return. The Bureau's decision was made in the hope of preventing Corrigan attempting a return flight across the Atlantic to New York. Although his license was revoked, Col. Monroe Johnson had said in an interview "The boy has made a hero of himself and we are tickled to death he got across. We are not going to be guilty of chilling the exploit by talking of punishment. Johnson also announced that Corrigan's plane would be brought home from Ireland in the Maritime Commission's steamer *Lehigh*.

Corrigan had no plan whatever of attempting a return crossing of the Atlantic by plane. He was already a hero in America and a huge civic welcome awaited him in Los Angeles. The Mayor of the city had cabled him "We are tremendously proud of you but please return the safest way - aviation needs you."

During his short stay in Ireland Douglas Corrigan was inundated with offers to make personal appearances in America, to join air circuses and there were film offers. He made no commitments to anyone before he sailed from Cobh for New York at the end of July 1938 having never deviated from his statement that he crossed the Atlantic in error.

At an 80th anniversary celebration for Douglas Corrigan in New York's Grand Hotel in the autumn of 1987, he was presented with a compass, a plaque and aviation maps by Harry Wagner, a director of the hotel. Although it was 49 years since that famous flight, Corrigan still stuck to his story.

"I followed the wrong end of the compass" he said to the gathering in his honour. "The only story I know is the one I've been telling and I've told it so many times I'm starting to believe it myself" he added with a chuckle. "Why should I change it now?"

He recalled that after repeatedly pestering the Federal Aviation Administration for permission to fly across the Atlantic they just told him to "Get lost".

"And I took at them at their word" he said. "I got in the clouds and I thought I had turned west. For 26 hours I flew in the clouds and rain. There were clouds below me so I couldn't see the water and clouds above me so I couldn't see the sun".

Was it a genuine mistake or did he turn his nose up at authority? On the surface it would appear that he made a genuine navigational error. Yet, it is strange that, throughout his long flight, he claimed to have never caught even a glimpse of the sun. Without any navigational charts of that part of the world, was it sheer luck that brought him over the most northern part of Ireland and later on very neatly pick out Baldonnell Aerodrome in Dublin? It is your guess!

THE FLIGHT THAT SHOULD NEVER HAVE BEEN MADE.

On Monday 6th October 1930 newspapers around the world carried banner headlines such as **WORLD'S WORST AIR DISASTER** etc. above detailed accounts of the crash of the British Airship R101 in the early hours of Sunday morning at Beauvais, 50 miles (80.5 km) north west of Paris, with the loss of 48 lives from the 54 that were on board.

Airships were not a new development. The first powered and manned airship was that constructed by the French Engineer Henri Giffard way back in 1852 which he piloted a distance of 16.8 miles (27 km). The decades that followed brought many changes and developments in airships but all of those belonged to classes known as non rigid and semi-rigid types. In the non rigid types the envelope shape was maintained solely by gas pressure. With the semi rigid airships, there was a rigid keel which allowed a lower gas pressure. A major drawback with those types was that a reduction in gas pressure caused a change in the airship envelope shape and if this continued, the forward end of the envelope would collapse making it extremely difficult to steer the airship.

The German Count Ferdinand Von Zeppelin was the first to make a serious attempt at developing a rigid airship in 1899. His first design was not very promising but it was a step in the right direction and he went on to more successful Zeppelin designs later. England, USA and France also began their developments in the early 1900s and of course the advent of the First World War brought airships more to the fore. The German Navy used them to a considerable extent in bombing raids on Britain.

The British used their airships for naval patrols which proved very effective in tracking U-Boats and passing information of their whereabouts to surface vessels of the Royal Navy. When the war ended, Britain's airship fleet was of poor calibre, the main additions just about to enter service were the R33 and R34. The latter made aviation history when between 2nd and 6th July 1919 it made the first airship Atlantic crossing.

The use of airships as long-range commercial transport was being considered for some time and was given very serious thought by the first British Labour Government under Ramsay McDonald. They allocated a sum of £1.3 million over a three year period to cover the cost of a limited amount of airship research, the construction of two large airships to be designated R100 and R101 respectively and the erection of mooring masts and new sheds both in the UK, Canada, Egypt and India.

By this time duralumin had been developed which was an ideal metal for the construction of an airship frame as it was extremely light relative to steel. A contract was signed for the construction of the R100 in October 1924 and the airship flew for the first time in 1929. Around the same time that the R100 was being built by the Airship Guarantee Co. which was a subsidiary of the Vickers Aircraft Co., the R101 was being constructed at the Air Ministry's Royal Airship Works at Cardington, Bedfordshire. It was a great boost to Bedford whose importance in the 18th and 19th centuries had declined after the first World War and the advent of the 1930s wasn't holding out much hope. Now there was great pride that the R101 project would restore some of Bedford's lost prestige while also providing badly needed employment. A great many of those involved, such as Colonel Richmond, the designer of the R101, the Director of Airship Development Wing Commander Colmore, the Assistant Director (Flying) Major Scott and the man who would eventually command the Airship Flt. Lieut. Irwin together with very many of the crew all lived in the area. Cardington where the R101 was being built was a suburb of Bedford.

Following completion, the airship's performance did not come up to expectations during the course of flight test. For example very strong vibrations were experienced when the engines were run at full speed and they could only be eliminated by reducing engine speed. Pre-flight trials showed that the disposable lift was only about half of what the government stated it would be in 1924. Under flying conditions the disposable lift tonnage would be further reduced, thus decreasing the fuel carrying capacity and putting restrictions on a long distance flight. The airship accordingly had to be lightened. Some of the cabins and toilets were taken out, a look out at the top of the airship was removed and among other things Cellon, which was a rather light celluloid, was used as a replacement for the Triplex glass windows.

Following the modifications a number of flight trials were made and the R101 (G-AHFA) made a number of flight trials. One of those trials was an endurance test of 48 hours but the airship returned to its moorings after 31 hours. There were more snags. An engine cut out during speed trials although the ship achieved 63 mph on one leg of the course. While moored to the mast in November 1929, the airship rode out a violent storm but excessive rolling of the ship caused most of its gas bags to move, often the valves opened.

In November 1929 the airship was returned to her shed for detailed examination and modification. The examination showed that all the gas bags had been holed, one having over one hundred holes in it. Following repairs, modifications were carried out to prevent the bags from chafing.

On 23rd June 1930 the R101 was again brought out of her hangar. This was always a delicate task for those who 'walked' the enormous ship and the event invariably attracted huge crowds. As she cleared the hangar the wind tilted her sideways and rent a tear of about 140 ft. (42m) long in her outer cover. This was repaired and the R101 was moored to her mast in readiness for flight trials prior to the airship's appearance at an RAF display at Hendon on 28th June

A further setback occurred the next day while the ship was moored when some rents again developed in the airship's cover. They were stitched and re-enforced with a rubber solution. At this time an engineer who was working on the R100 was on a visit to Cardington. He was later to say regarding the repair "The effect was to make the cover so flaky you could put your finger through it." That engineer later became a famous writer - Nevil Shute. Following repair the R101 had a short trial flight on 26th June.

On the following day a dress rehearsal test flight for Hendon was undertaken which was far from reassuring. The performance of the airship was, to say the least, disturbing. She pitched quite a lot and the crew were obliged to jettison 9 tons of ballast. Captain Irwin was anything but satisfied with his ship and he filed his report accordingly.

At the RAF Display at Hendon the airship behaved reasonably well. However, on the way back the reverse occurred. Having flown over Luton she went into some short dives which caused the coxswain to remark to Captain Meagher, a very experienced flyer and who was acting as First Officer, that it was as much as he could do to keep the ship up. Around 10 tons of ballast and fuel had to be jettisoned before eventually the R101 returned to normal flying and she was later moored to her mast. On the following day the airship was taken into her shed for a thorough inspection and further necessary modifications .

It was the British Government's intention that the R101 would travel to India with Lord Thompson, Secretary of State for Air on board at the end of September. So considerable pressure was being brought to bear to have all the necessary modifications and flight trials completed as time was running out. There was now less than twelve weeks left.

The inspection that followed showed that 15 of the airship's 16 gas bags, had been badly holed, despite an earlier fitting of padding to prevent chafing. The gas bag material was made from oxen intestines, then felt to be the best to securely contain the five and a half million cubic feet of gas being carried. Number 11 bag had actually 103 holes in it.

    This continuous piercing of the R101 gas bags had been the subject of several reports by Captain Irwin but neither the airship's designer Col. Richmond nor Wing Commander Colmore who was Director of Airship Development was able to suggest any alternative other than more padding to prevent chafing.

    Even at this early stage of the airship's existence there was tension building up between many of the decision makers. An aviation inspector named Wade was not too happy in his dealings with the designer and with his remedies regarding chafing etc. He was not making great headway either with Colmore so he decided to go over both their heads. He wrote a confidential letter to the Air Ministry directed at either the Air Member for Supply and Research or Lord Thompson, the Secretary of State, in which he said

    "Until this matter is seriously taken in hand and remedied, I cannot permit you the extension of the present "permit to fly" or the issue of any further permit or certificate."

    The letter apparently never got to them but was returned to Colmore who directed McWade to go ahead with more padding. Among the other modification decided on was the lengthening of the airship by the insertion of an additional gas bay to give her extra lift. By so doing it brought her overall length to 777 ft. (237 m.) This meant that the R101 would have to be sliced in two. Time was running out, the 22nd September having been fixed as the deadline for all modifications so that flight trials could immediately commence.

    However, an order was issued by the Air Ministry that the R101 be not tampered with in that way lest the R100 was not ready to go to Canada and the R101 would have to go in her place. This was amazing logic considering the behaviour of the R101 up to then. The R100 did in fact go to Canada and returned safely.

    On 21st July Lord Thompson allowed the order, which had been issued by Sir John Higgins, Air Member for Supply and Research, to be cancelled thus giving the go ahead for the necessary major modification. It presented a daunting task that would leave just about a week in which trials could take place to satisfy safety regulations before the R101 would fly from Cardington to Ismalia and from there to Karachi. Work went ahead at full speed but despite the efforts of all concerned the 25th September was the earliest date the airship would be ready to fly. Taking flight trials into consideration, Saturday 4th October was now fixed as the departure date to India. This was assuming of course that the flight trials would be satisfactory thus allowing a Certificate of Air Worthiness for the R101 to be issued.

In mid September, Air Vice-Marshall H.T.C. Dowding replaced Sir John Higgins as Air Member for Supply and Research. Later to become Lord Dowding, he knew little or nothing about airships and had never flown in one. He was an aeroplane man of distinction and his inspiring leadership in the RAF was to play a major part in winning the Battle of Britain in the Second World War. However in his new post in September 1930, Dowding relied very much on his adviser on airships, Wing Commander Colmore.

To date the R101 had only two periods of flying time in her rather brief existence. Most of this time (102 hours in all) was done in 1929 before alterations had been made to her gas bags. The remainder was absorbed in two short flights following the letting out of the bags. Those two flights had not proved satisfactory.

Colmore's experience with airships led him to believe that 150 hours flying time should be undertaken before an intercontinental flight took place and speed and endurance trials would have to be encompassed in that period. It depended of course how that figure would be interpreted. Where the R101 was concerned. there were two ways. One was that the airship had already flown 102 hours leaving a balance of only 48 hours for endurance and speed tests.

Under normal circumstances that would be a reasonable conclusion to come to but one was not dealing with normal circumstances. The 102 hours flying time already undertaken by the airship was done mainly in 1929 and since then gas bag modifications had been made and a new section fitted. Taking all this into consideration one would deem it prudent that far more than 48 hours extra flying time be undertaken before a Certificate of Air Worthiness could be issued. Sadly this was not to be and in fact the 48 hours flying time still outstanding were not even adhered to.

Lord Dowding was anxious to see things for himself and as a 24 hour flight trial had been arranged, he decided to go on it. As it transpired, because of other commitments he could only manage 17 hours. The flight proved very little. The weather was extremely calm and dry and there was no test of the airship's behaviour in rough weather or that the previous faults had now been cleared. In fact, after a few hours flying, the airship's starboard forward reversing engine began

The R101 Imperial War Museum, London

to leak and a new washer had to be fitted. As a precaution, the engine was shut down and the port engine was reduced to half speed. The designer Richmond together with Colmore were happy with the extra lift that the newly fitted central gas bag had given the R101. Dowding was impressed with the airship's behaviour but really - he was an aeroplane man!

Following the reconstruction of the R101 an interim engineering report had been given to the Air Ministry stating that the aerodynamic design and structure conformed to the requirements laid down. However, the Aeronautical Inspection Department following its examination was critical of the gas bags and the new methods undertaken to prevent them from being chafed and holed. Colmore, Dowding's adviser, felt that the criticisms of the Aeronautical Inspection Department would be put right before departure on 4th October.

It was now 2nd October when the R101 returned to her mooring mast. Dowding and Colmore reported to Lord Thompson but before they did so the former had said to Colmore "You are my adviser and whatever line you take with the Secretary of State I shall back you up.". A Certificate of Air Worthiness was drawn up following the meeting and this was signed by the Deputy Director of Civil Aviation. That certificate was issued even though the R101 had not undertaken or passed her high speed trials. Dowding felt that such trials could be undertaken soon after the airship left her mooring on her voyage to India. This was not very practical as it could prove an embarrassing situation for Captain Irwin should the speed trials prove unsatisfactory and he was obliged to return, especially with all the top aviation men on board.

Irwin was far from happy and was extremely worried. He had thought of refusing to go and resign his short service commission. This wouldn't have achieved anything as Lord Thompson and the Air Ministry were adamant that the flight would take place Should Irwin refuse to command the R101, it would be commanded by his great friend Squadron Leader Booth a regular officer who would be obliged to take control. Thompson had heard of Irwin's misgivings and accused him of obstructionist tactics. Dowding from the advise given him was happy enough.

One senior aviation official was very worried about the trip and had premonition of pending disaster. He was Sir Sefton Brackner, Director of Civil Aviation. There had been many stories forecasting doom for the R101 floating about Cardington. Emilie Hinchliffe the widow of noted airman Captain Hinchliffe, who in March 1928 disappeared over the Atlantic together with the shipping heiress the Hon. Elsie Mackay in an East-West Atlantic crossing attempt, had visited a well known medium named Eileen Garrett in an attempt to contact her dead husband. Mrs. Hinchliffe claimed that she had been told by him through the medium that the R101 was doomed and to warn his old friend Squadron Leader Johnson, the navigator of the forthcoming flight to India not to make the trip. Johnson of course laughed this off. Mrs. Hinchliffe then got the famous author Sir Arthur Conan Doyle to vouch for the medium's integrity but again he was not impressed.

Sir Sefton Brackner had been told by a lady that the R101 would crash in flames. That lady was the medium Eileen Garrett. Some time previously, Miss Auriol Lee a lady friend of his who held him in very high regard asked an astrologer to look at Sir Sefton's horoscope and was told that he would meet death by fire.

Whatever was being told by the medium or what was forecast in the stars, Brackner had heard of the many adverse comments made by officers who had already flown in the R101. He went and saw Thompson and as diplomatically as he could told him that in some circles the wisdom of undertaking such a trip at the present time without adequate and satisfactory flight trials was being questioned. The Secretary of State for Air did not relent in any way and even politely suggested that if he did not wish to make the voyage there were many others who would eagerly take his place. Brackner did not withdraw his name.

There were others too who were not confident of the R101. The editor of *The Aeroplane* felt that she would get no further than Ismalia. Colmore was making the flight but felt that additional mooring masts along the way should be provided, in case the airship had to land for fuel other than at Ismalia.

The stage was now set for the flight to India in an airship that did not appear to have met the requirements which one would have felt reasonable for such a voyage. The Captain of the flight was Irishman Flt. Lieut. Irwin who was an experienced and very thorough and responsible officer. Irwin had spent all his flying service with the RAF in airships and was a very fine flyer. His several planned flight trials for the R101 following the insertion of a new centre gas bag etc. were turned down. Yet he had been presented with a Certificate of Airworthiness on the morning of the flight.

The question of captaincy of the R101 was a thorny one. Officially Flt. Lieut. Irwin was Captain even though he was one of the most junior ranking officers on board. Scott, who was among the passengers listed, was much more senior in rank and in airship hierarchy. He didn't like this state of affairs and let it be known. Both had sought clarification regarding the position of who was in charge of the airship. The Air Ministry had a delicate problem on their hands and sought to resolve it in a typical civil service manner. They wore two hats. While to the outside world Irwin was Captain, they had a communiqué drafted for release to the world press when the airship had slipped her moorings that evening and which read as follows:-

"*Airship R101 left the mooring tower at Cardington at .... hours GMT, 4th October on the first stage of her flight to India. The flight is being carried out under the direction of Major G.H..D. Scott, CBE, AFC, Assistant Director in Charge of Airship Flying.*"

As to the possibility of being able to carry out the necessary high speed trials shortly after take off, such an idea was firmly dashed because the weather deteriorated. Just after 1500 hours a weather report was received stating that the occluded front which was over France this morning had passed eastwards "*while a trough of low pressure off western Ireland is spreading quickly. Cloud is increasing to ten tenths and falling to 1000 feet. Rain will spread from the west, probably reaching Cardington tonight.* "

As the huge craft awaited take off, there was great activity while the passengers and their luggage were settled in. Lord Thompson's baggage was largest of all despite the fact that to keep the weight as low as possible each crew member was only allowed to take the very minimum. As well as nine pieces of luggage, he had brought cases of champagne, a roll of carpet "to impress the Egyptian King" and a specially fired Wedgwood dinner service to grace the dinner table when his Majesty dined on board. In all, the R101 carried a crew of 42 and 12 passengers. Captain of the airship was Flt. Lieut. H.C. Irwin, AFC. RAF. The First Officer was Lt. Commander N.G. Atherstone, AFC. (R.N. retd.) and the Navigator was Squadron Leader E.L. Johnson, AFC. RAF. who had flown in that capacity on the R100 to Canada. Among the passengers were Lord Thompson, Secretary of State for Air; Air Vice-Marshal Sir Sefton Brackner, Director of Civil Aviation; Wing Commander Colmore, Director of Airship Development and Major G.H. Scott, Asst. Director of Airship Development (Flying).

It is extremely questionable why the R101 left Cardington at all on Saturday 4th October at 6.30 pm given the unfavourable weather reports from the Continent. It was not Captain Irwin who made that final decision to take off, it was Scott.

When the airship left her mooring mast the weather was not good and heavy rain was falling. The cloud ceiling was around 1500 ft. On the first leg of the flight the airship was carrying 25 tons of fuel. It was not a very auspicious start. Due to the adverse weather the R101 rolled and pitched somewhat. In less than two hours she was over London and to avoid the worst of the weather in parts of Europe, course was set for the Mediterranean. Such a flight path would bring the airship over Paris, Tours and Narbonne. The R101 reached the coast at Hastings at 9.35 pm and was then flying at a level of around 1000 feet. At that stage of the flight one of the airship's engines was temporarily out of action. It reached the French coast an hour and fifty minutes later, still achieving no great height. Many who saw the ship passing on its way towards the capital felt that she was flying extremely low, some stating that she was flying little higher than the buildings.

The last radio message to be picked up from the R101 was that by Le Bourget Aerodrome on Sunday at 1.50 am. This did not give an indication that anything was amiss and the message added "the passengers, after an excellent meal and after enjoying a number of cigars are getting ready to go to bed." The airship's position as given by Le Bourget was "1 Km. north of Beauvais at 1.52 am, flying at a height of 1000 ft."

At 2 am the R101 passed over Beauvais where eye witnesses remarked on the low level of its flight. Shortly afterwards it plummeted to earth and burst into flames. Of the 54 on board only 8 survived the crash, two of those later died. Among the dead were Thompson, Brackner, Colmore, Scott and the Captain, Irwin. It was an appalling tragedy that should never have happened.

Before departure, Lord Thompson in an interview with the press had said "I am starting off with great confidence. It seems that conditions are quite satisfactory." One of the survivor's of the terrible tragedy was foreman engineer H.J. Leech. He gave a very vivid account of what happened.

"The night watch composed of 12 men was at its post. I personally was in the navigation cabin. The trip was proceeding without incident until just before we arrived over Beauvais when we were overtaken by a terrible storm with squalls of wind and rain and violent whirlwind eddies. Nevertheless, I had confidence in the solidity of the airship and pushed on unconcerned. However, over Beauvais at about 1.40 I had the impression that the airship was pitching dangerously. We were making headway - very slowly it is true - but still we were making headway. Le Bourget communicated our position to us at 1.43 am as two Kilometres south of Beauvais. I had no idea at what height I was navigating and I had to resist the force of the wind with all my might. Then the airship began to feel the weight of the rain and responded badly to the motors which I ordered to be driven full speed ahead in order to try and obtain altitude.

Then suddenly disaster occurred. The nose of the airship, after dipping twice, struck the ground violently on the top of a small hill. A terrible explosion followed and everything began to blaze. Hurling myself against the cabin wall and smashing it with anything that came to hand, I managed to make an opening and dashed through the flames head first.

The hill is 780 feet above sea level. The forward engine bars were crushed under the hull which in turn collapsed and buried the cabins in wreckage. I counted a space of about a second between the crash and the explosion which was followed by a roar of flame and the hull burned from end to end. My idea is that the explosion was due to the breakage of electric wires which caused the flash."

Mr. Leech having got himself clear heard a voice calling from the wreckage for help and in a fine display of bravery went back and extricated the wireless operator whose name was Disley.

Later, the survivors together with the charred remains of those who lost their lives were brought to England by the destroyers *Tribune* and *Tempest*. Cardington and Bedford were devastated by the tragedy. It was to Bedford that the funeral train carrying the coffins proceeded from Euston. Most of those who died were buried in Cardington churchyard where a huge grave was dug to accommodate the coffins.

Within a couple of weeks an Inquiry into the crash was opened . It was presided over by Sir John Simon. Representing the Government, who looked upon the proceedings with grave importance, were the Attorney General and Sir Stafford Cripps, the newly-appointed Solicitor General. The assessors were Colonel Moore-Brabazon holder of the first pilot's licence issued in England in 1909 and Professor Inglis of the Engineering Department at Cambridge.

If one were looking for a 'fall guy' in relation to the crash of the R101, it would be logical to point the finger at her Captain, Flight Lieut. Irwin. With him, so to speak, the buck stopped even though on board that airship he was not alone out-ranked but out-commanded, despite report after report that he had filed expressing his dissatisfaction with the airship's performances before that fatal flight. Lest there be any

Wreckage of the R101

Imperial War Museum, London

move to put the blame on him and destroy his good name, his wife Mrs. Irwin was represented at the Inquiry by Major Tweed. Of all the families who had been bereaved by the tragedy, she alone had legal representation.

As is often the case where an Inquiry is set up, documents have a habit of going 'missing', especially if they would prove in any way incriminating. When Sir John Simon, President of the Inquiry asked very early on in the proceedings who authorised the shortening of the final test flight to 16 hours, no one could give him an answer. When he sought the test flight report, the Attorney General told him that no such report existed. It was not a very auspicious beginning to get to the bottom of the cause of the fatal crash. In fact many documents had mysteriously gone 'missing'

When pressed about the absence of the flight trial report, the Attorney General was forced to admit that it was "very remarkable." One very serious aspect of the Inquiry so far was that a most important report made by Captain Irwin on the gas bags problem had also gone 'missing'. The Inquiry President directed a thorough search be made at Cardington for the relevant documents but none were found. Then for some mysterious reason a number of them were found including that made by Captain Irwin on the gas bags.

Also among the documents now 'found' was a letter written by Mr. W. F. McWade, Inspector of Cardington Maintenance to the Air Ministry Secretary pointing out that the number of holes which had occurred in the gas bags "amounts to thousands". When Sir John asked why those very important documents were not produced before, there was no reply. He was evidently doing his utmost to get into the open any moves that had been made towards a 'cover up' and he asked Inspector McWade what he would have done had he the final say in issuing a Certificate of Air Worthiness. McWade replied "I'm afraid, they would never have got it."

As the weeks of the Inquiry went by it was clear that Captain Irwin was blameless where the crash was concerned and at one stage during the proceedings Sir John said in a reference to Irwin "I may say that I am personally completely satisfied that there is no possible grounds to cast reflection on him".

After four months a report of the Inquiry was published stating that the most likely cause of the R1O1 crash was that a gas bag forward in the airship suddenly burst, bringing the airship's nose down too rapidly to be controlled. Neither Lord Thompson, the Air Ministry or the Royal Airship Works came in for any criticism. Neither of course was there any criticism levelled at Flight Lieut. Irwin as Captain of the R101. However, he certainly wasn't commended for the responsible reports, often adverse, that he had sent to those in authority and which if heeded may have averted the awful disaster. To do so could easily have tipped the balance of blame on to Lord Thompson and his Air Ministry for allowing the airship begin the flight to India.

In fairness to Sir John Simon, he went out of his way to ensure that Captain Irwin's name would not be unjustly tarnished and some time after the Inquiry had ended wrote:

"One of the things about which I was particularly pleased is that Irwin's record and reputation stood out beyond challenge. Poor Mrs. Irwin must get such consolation as this affords."

No definite cause was found for the crash of the R101, One thing is clear. The R101 had never sufficient and satisfactory flight trials after the addition of a centre gas bag and modifications to prevent the gas bags from chafing. There would also appear to have been too much concern at having the airship ready to leave on 4th October as if an entire aviation programme depended on such an operation.

Finally, a Certificate of Air Worthiness is specifically what it says and one questions the validity of issuing such a certificate under the circumstances. It was well known that grave doubts had been expressed by many whose observations were not made lightly. But they were like voices crying in the wilderness. Today there is little to be gained in apportioning blame.

Suffice to say that the end did not justify the means and the flight of the R101 from Cardington to India which began on Saturday 4th October 1930 should not have been undertaken.

AMELIA EARHART

Her Flights and Mysterious Disappearance.

When Amelia Earhart and her navigator Fred Noonan set out on 1st June 1937 on a west to east flight around the world just north of the Equator, it was never anticipated that both flyers would disappear without trace, sparking off one of the greatest search operations by the US defence services at that time. Was it just a genuine crash? or were the two aviators captured by the Japanese having come down on one of the islands they held. American government sources added to the mystery by their reticence and being less than frank with their answers and explanations to the many questions surrounding the incident. Today, over half a century later there has been no final answer.

Before we look at the story behind that fatal flight let us look back on the flying career of one of the greatest women aviators in the world. Amelia Earhart was born on 24th July 1898 in Kansas City to Edward Earhart, a struggling lawyer and his wife Amy whose father was a judge. The latter would never accept that Earhart was a suitable choice for his daughter and in this he was later proved to be right. It was nine years before Amelia's father got himself a permanent job with the railroad claims section in Des Moines, Iowa to where the family moved. He became an alcoholic. He lost his job and eventually could only find work as a law clerk with the Great Northern Railway in St. Paul's, Minnesota where the family again moved to.

It was a disturbing time for young Amelia and her sister Muriel. The drinking continued and the family split for some time, Edward Earhart moved back home to Kansas City while the rest of the family went to Chicago. It was during the First World War in 1917 that Amelia first became interested in aviation. She was spending a Christmas holiday with her sister in Toronto and saw the wounded being brought home from the Western Front. She asked her mother to allow her stay on and become a nurse's aid. There Amelia met Captain Spalding of the Royal Flying Corps who invited her to the airfield to see him flying. Very soon she was captivated by aviation. However, she had planned to become a doctor but domestic problems with her parents did not allow this.

While in California in 1920 Amelia Earhart began taking flying lessons and eventually passed her flying tests. Her parents, especially her mother and also her sister, bought her an aeroplane for her 24th birthday in July 1922. The plane was a Kinner Canary bi-plane.

Two years later, in the Summer of 1924, Amelia Earhart was obliged to give up flying for the time being. Her parents had agreed to a divorce and Mrs. Earhart moved with her two daughters to Boston. Amelia reluctantly sold her plane and bought a touring car instead to bring her mother and sister east. Over the next few years she became both a student and teacher and after six years could boast that she had held twenty eight different jobs. She wasn't cut out for teaching and went into social work which she liked. The call of the sky however was always there.

While employed as a social worker she received a letter from William Kinner a plane manufacturer in Los Angeles from whom her first plane had been bought asking if she could assist him in getting an agent to sell his planes in the east. While investigating this she met an architect named Harold T. Dennison who had an interest in aviation He had land on which he had constructed a basic landing strip and had plans to form a company that would build a commercial airport. On learning that Amelia was keen on aviation, he asked her to become one of the board of directors of the company. This she did and began flying again, demonstrating the Kinner plane to those interested. Although still employed in social work she began making plans to form an organisation of women flyers and contacted many of them.

Before this could be put on a sound footing she received a phone call one day in April 1928 from a Captain Hilton Railey whom she had never heard of previously and who, out of the blue, asked her if she would fly an aeroplane across the Atlantic in the cause of aviation. The opportunity was too good to turn down and she agreed. She was later to learn that the proposition was not as clear cut as she was led to believe at first. Captain Railey however was a very respectable man and head of a large public relations company and was a friend of the publisher George Palmer Putnam.

Putnam had heard that a wealthy woman had bought a tri-motored Fokker Friendship from the aviator Commander Bird and was attempting to fly the Atlantic in it. Railey obtained the inside story on this plan for Putnam and they managed to contact the sponsors who gave them the job of arranging the flight which was to be surrounded with great publicity. The wealthy lady was Mrs. Frederick Guest of London. Her husband had served in the Air Ministry during the First World War and she hoped by this public relations exercise to consolidate relations between America and England.

Mrs. Guest's family talked her out of flying the plane and a search was made to find a woman pilot to personify what she had in mind. Captain Railey was told of a young social worker named Amelia Earhart. He interviewed her and was very pleased with the result. He found her bright, intelligent and highly articulate. She had by this time 500 flying hours clocked up but had no long distance flying experience. She also could not fly by instruments and had never flown anything other than a single- engine plane.

On the surface, the entire project was bizarre but it emerged that Amelia Earhart would not be flying the plane, there would be a male pilot and a mechanic on board. She was also to learn that the pilot would be paid $20,000, the mechanic $5,000, while she herself would get no cash payment, only the honour and glory of becoming the first woman to cross the Atlantic in a plane. She still went along with the idea and eventually met the other two crew members. Bill Stults was a highly talented flyer whom she later discovered was an alcoholic but once he was behind the controls an excellent pilot. The mechanic was a Texan named Lou Gordon.

The plane had been modified and fitted with pontoons instead of wheels. On 4th June 1928 with favourable weather reports the *Friendship* lifted off from the water near East Boston and eventually landed in Halifax, having attracted much press publicity. Next morning they set off for Trepassey Bay, Newfoundland, the last landfall before making the arduous Atlantic crossing. There they landed safely, but due to weather conditions were forced to stay for the next thirteen days.

Amelia Earhart was later to recall that during this time of waiting Stults began drinking heavily and so casting grave doubts in her mind of his capability to fly the Atlantic. However, when the day of departure came on 17th June she and the mechanic got him on board where he quickly sobered up and took off shortly after 11 am. Stultz was now happy and in full control. Twenty hours and forty minutes later the plane landed at Burry Point in Wales and Amelia Earhart had become the first woman to fly the Atlantic in a plane but had not piloted it.

A huge crowd of well-wishers turned out to greet the aviators once the news had broken. To commemorate that occasion, a monument 18 ft. high was later erected at Burry Point in honour of the *Friendship* crew, Amelia Earhart being saluted as being "the first woman to fly over the Atlantic."

With the sponsor of the flight awaiting them in Southampton, the plane eventually took off for there. On this short leg of the flight, Stultz allowed Amelia to fly the plane. On landing at Southampton they were again warmly greeted, while in London the welcome was much more exuberant. The publicity surrounding the flight once it had taken off from Newfoundland was immense and in a very short time Amelia Earhart had become a celebrity. For two weeks London opened its heart to her. At times she found it very embarrassing especially when she was being given the distinction of flying the plane across the Atlantic. To her great credit Amelia Earhart took every opportunity to clarify the situation. She was later to say to Captain Railey "I am a false heroine and that makes me feel guilty. Some day I will redeem my self respect. I can't live without it."

While in London she took time off to go and meet the Irish aviator Lady Mary Heath whom she greatly admired. Amelia got a few hours flying in Lady Heath's Avian Moth which she kept at Croydon Aerodrome. In fact before leaving she bought the plane and had it later shipped to America.

After a hectic two weeks in England Amelia Earhart and her crew returned by ship to America where they received an enormous welcome including the traditional Broadway parade. She was now hot property and a sponsor's dream. George Putnam who was later to marry Amelia acted as her manager and advised her on endorsing products, giving lectures, writing articles etc. Within a few months she had earned about $50,000. Putnam was a publisher and quickly got Amelia to write a book about the flight which was called *20 Hrs. 40 Mins.* It was written by her in Putnam's suburban home in New York where he lived with his third wife Dorothy. Ironically Amelia dedicated the book to her. The Putnams separated in 1929 and in the following year got divorced.

Amelia Earhart had now arrived on the aviation scene fully determined to become a great aviator in her own right and not just an endorser of products or a writer. She had become financially independent in a very short time. In November 1929 she and another fine woman aviator Ruth Nichols formed a women's flying organisation at Curtiss Airport, Long Island. Amelia became its first president.

Her one great aim at this time was to make a solo flight across the Atlantic but meanwhile she was looking at the future of commercial aviation and how people could be encouraged to fly. She was asked in 1930 to join a newly established airline, the New York, Philadelphia and Washington Airway to highlight the advantages of air travel. She of course continued with her own flying and in early April 1931 while flying an auto-giro set an altitude record of 18,451 ft.

From the time that George Putnam first met Amelia Earhart in 1928 he was smitten by her and proposed marriage on several occasions. She turned him down, a setback and a surprise to the often aggressive high-powered business man who was used to getting his own way. Amelia was a very independent minded lady and if she was to marry George Putnam it would be on equal terms with him. She had inside knowledge on how marriages failed, her own parents being a case in point. Putnam's persistence eventually succeeded and she married him on 7th February 1931. The marriage worked well between two strong-willed people and in fairness to him, because of his great business acumen he successfully promoted the commercial side of his wife's aviation work without infringing on her independence.

In the Spring of 1932, she and her husband invited a famous Norwegian aviator Bernt Balchen to lunch at their New York home. Amelia held him in the highest regard. He was no publicity seeker and his advice was always sound. She told him that she wanted to fly the Atlantic and asked him would he help her if he thought she was capable of doing it. He agreed.

Bernt Balchen would act as technical adviser for the flight. There would be a complete veil of secrecy thrown over the idea, the take off which was planned from Harbour Grace, Newfoundland would not be publicised. Amelia Earhart wanted the whole affair to be kept in a low key. The plane she would use would be her own Lockheed Vega monoplane.

Work went ahead on the aircraft at Teterboro Airport, New Jersey. A new 500 H.P. Wasp Engine was fitted and extra fuel tanks were added to the wings and one in the cabin. Amelia took a detailed course in Meteorology at the New York Weather Bureau. Balchen saw to it that nothing was left to chance and the plan was that when the weather was favourable he, with a mechanic and Amelia on board, would fly to St. John's and from there to Harbour Grace from where she would begin her solo crossing.

Favourable weather gave them the go ahead on 20th May and the three took off from Teterboro Aerodrome to St. John's where they overnighted and then flew on to Harbour Grace. Amelia Earhart was carrying the minimum of luggage, certainly she had no evening wear should she successfully cross the Atlantic and be inundated with invitations to receptions, dinners etc. at the other side. For food she had packed a thermos of soup and a can of tomato juice which she had punctured and inserted a straw. Before the flight she had commented that "a pilot whose land plane falls into the Atlantic is not consoled by caviar sandwiches."

When she took off alone from Harbour Grace there was a forecast of

Amelia Earhart with Mr. McCallion in Derry, N. Ireland, the first person to greet her on the 21st May 1932 following her solo Atlantic crossing.
Courtesy Short Bros. plc. Belfast

reasonable weather. However, she encountered some very disturbing situations. An altimeter failed, the plane's wings began to ice up and at one stage because of this it went into a spin and dropped vertically by 3000 feet. Later on the weather settled down and her Lockheed Vega behaved very well except for a strong vibration coming from an exhaust manifold. She eventually picked up Donegal and following a railway track came over Derry. She circled the city but was unable to find an aerodrome and was obliged to put the plane down in a meadow.

Amelia Earhart had done it and created history by becoming the first woman to fly solo across the Atlantic. Her time was 15 Hrs. 18 Mins. reducing the time it took the *Friendship* to cross by 5 hours. The world press quickly got the news and soon she was at the centre of enormous adulation. London, Paris, Rome and of course the US were at her feet yet she remained very modest despite all the publicity and in response to President Hoover's speech of congratulation said that "the appreciation of the deed is out of proportion to the deed itself."

Among her friends were President and Mrs. Roosevelt together with the Lindberghs and many aviation greats. She continued to fly despite a very busy schedule. In August 1932, just a few months after her record-breaking solo Atlantic crossing, she established the women's nonstop transcontinental speed record, between Los Angeles and Newark in 19 Hrs. 5 Mins. In July of the following year she broke her own record between the two locations by an hour and fifty eight minutes.

In 1934 Amelia Earhart planned to fly from Hawaii to California, the longest flight she would make to date and which would necessitate crossing over 2,400 miles (3862 km) of the Pacific Ocean. There was a prize of $10,000 which was sponsored by Hawaiian business men. She engaged as her technical adviser for the flight an excellent airman named Paul Mantz who had been a stunt flyer in Hollywood films such as "Men with Wings" and "Hell's Angels". Her plans were made in secrecy.

When her proposed flight eventually became known there was some political controversy over US sugar tariff concessions and the sponsors began to have second thoughts. Amelia addressed them and stated that regardless of the adverse publicity which they appeared to be concerned with, she would fly to California from Hawaii with or without their support. The sponsors agreed to go ahead with the prize money for the flight. At 4.30 pm on 11th January 1935 she took off in a new Lockheed Vega and 18 Hrs. 15 Mins. later she landed at Bay Farm Island Airport near Oakland and so became the first aviator ever to have successfully

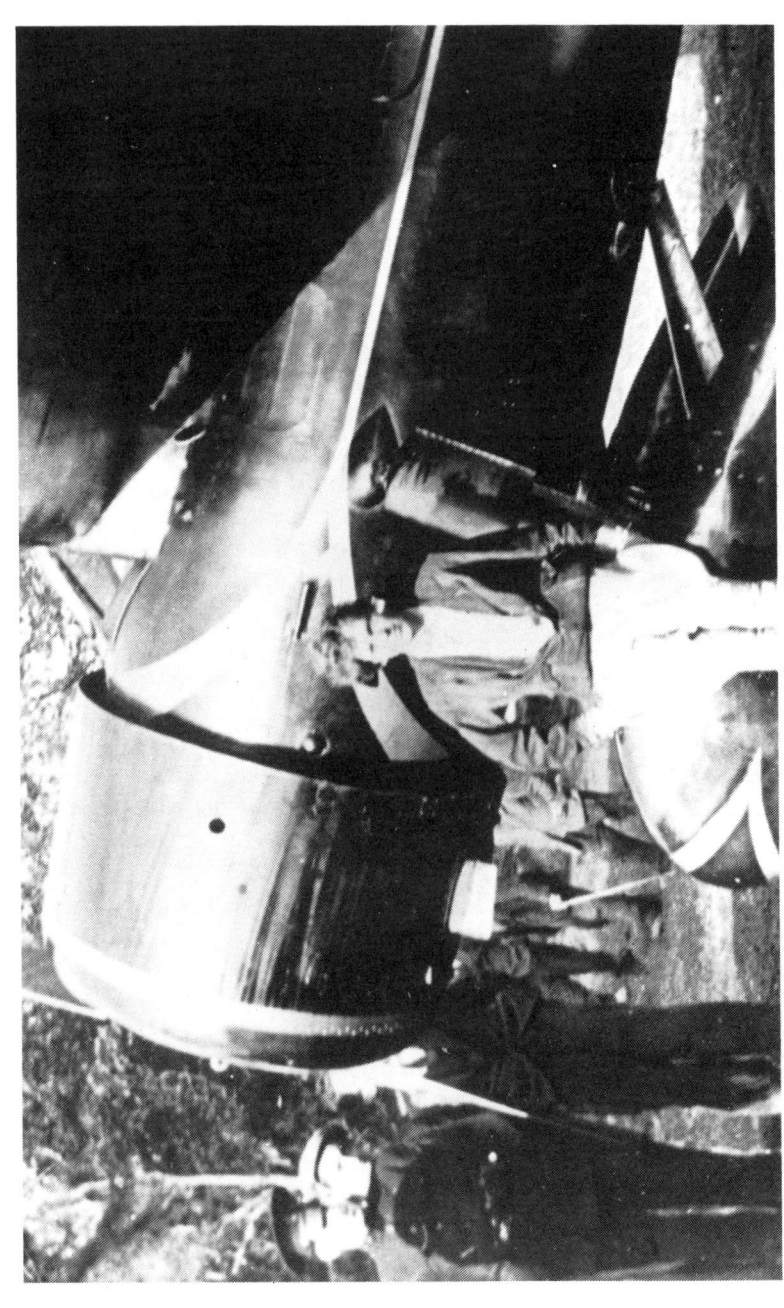

Amelia Earhart pictured by the fuselage of her Lockheed Vega plane in which she made her solo Atlantic crossing.
Courtesy Short Bros. plc. Belfast

made such a flight. The flight was highly publicised and about ten thousand people turned out to welcome her when she landed.

In April 1935 she made the first solo flight from Los Angeles to Mexico City. The idea for the flight came from the President of Mexico who had invited her to come and meet him. On 8th May Amelia successfully made the first solo flight from Mexico City to Newark. Later she wrote - "One's imagination toyed with the thought of what would happen if the single engine of the Lockheed Vega should conk ... and I promised myself that any further over-ocean flying would be attempted in a plane with more than one motor."

Amelia Earhart was very much a feminist and always promoted the equality of women inside and outside marriage. She was invited to join the faculty of Perdue University to lecture women students not only on careers as aviators but in many other walks of life also. She accepted the invitation and began in the Autumn of 1935 and was very successful.

From her University post there emerged a situation which would lead to her last flight. The University set up the Amelia Earhart Fund, its principal function being to purchase a new aeroplane which would be used by her as a type of flying laboratory where, among other things, the effects of flying on people could be studied. Amelia was delighted with the idea and hoped to "produce practical results for the future of commercial flying and for the women who may want to fly tomorrow's planes". The new aeroplane was the most sophisticated commercial machine of its kind at the time, a Lockheed 10-E Electra capable of carrying ten passengers. It was twin-engined and had dual controls with a range of over 4,000 miles (6,437 km) and a fuel capacity of 1200 gallons.

She took delivery of the aeroplane on 24th July 1936. By then she had already made the decision to fly around the world just north of the equator, a distance of about 27,000 miles (43,451 km), far greater than had ever been flown before. Her original idea was that she would fly and east-west course. It was an undertaking that required enormous planning and would also be costly. Many experts had to be employed, their knowledge vital for the success of the flight. Her technical adviser was again Paul Mantz. She undertook intensive training by him, especially in instrument flying.

Eventually all the necessary technical details had been worked out and refuelling and maintenance schedules arranged. The navigator she chose was Captain Harry Manning, a close personal friend. Also on board for the first leg of the flight to Honolulu would be Fred Noonan, a highly experienced Pan American Airways pilot. She had the full

The Lockheed Vega Monoplane in which Amelia Earhart made her solo Atlantic crossing.
Courtesy Short Bros. plc. Belfast

support of her husband George Putnam in this extremely long flight but many of her friends and aviation colleagues felt that she should not undertake the flight, the risks being too great. She understood fully their concern but nothing would change her mind.

On 17th March 1937 Amelia Earhart took off from Oakland on her round-the-world flight, the first stop being Honolulu. Paul Mantz joined this leg as he wished to visit his fiancee in Hawaii. He acted as co-pilot on the 2,410 miles (3,878 km) leg. On the following day they landed safely in Hawaii. Later he was to write that it was he who landed the plane at Amelia's request and that she looked "groggy" at the end of the flight. Mantz left the flight there and Amelia and her crew flew to Luke Field close to Pearl Harbour from where her next take off was scheduled. Due to an approaching storm her departure was delayed until 20th March, giving her time to rest.

Early on that morning, Amelia Earhart with her crew of two, Fred Noonan and Captain Manning took off for Howland Island in the centre of the Pacific. The island was owned by the US who had built an emergency landing strip there for her. The take off was not very spectacular. The plane in fact never lifted off but swung to the left and Amelia, despite her best efforts, could not prevent it from swinging about on the runway, shearing off one landing wheel and damaging a wing. Luckily no one was hurt. But the plane was badly damaged and had to be taken to the Lockheed factory in Burbank for repairs.

Meanwhile Captain Harry Manning opted out of the round-the-world flight attempt. The crash at Luke Field may have given him second thoughts on the safety of the undertaking. Fred Noonan agreed to go ahead with Amelia on the entire flight. He was vastly experienced as an aviator and was also a first class navigator. He had been a Pan American pilot and at a later stage had been manager of the Pan American airport at Port-au-Prince.

It took two months to get the Lockheed Electra in flying order again and Amelia tested the plane to her satisfaction. There was now a change of plan and it was decided to fly around the world in a west-east direction instead of east-west. She flew the plane to Oakland where Fred Noonan joined her and on 20th May they left for Miami. They touched down at Tucson Arizona for fuel. On board for this part of the flight were her husband and a mechanic. However, near disaster again struck when the plane had been refuelled. The left engine, following a backfire, went on fire. Luckily the ground crew extinguished it. With only slight damage, she and Noonan were able to take off next day for Miami which they reached on 23rd May. Considerable delay was experienced there because Paul Mantz, the technical adviser, wanted a complete check carried out on the plane before they set off on their arduous flight.

On 1st June 1937, Amelia Earhart, now almost 39 years of age, bid farewell to her husband at Miami aerodrome as she sat in the cockpit of her Lockheed Electra with co-pilot Fred Noonan. At 5.56 am the plane took off in excellent weather and headed for San Juan in Puerto Rico. In the days ahead they would fly from there to Carpito in Venezuela and then to Fortaleza in Brazil where Pan American had excellent facilities. The plane was overhauled there and on 6th June they flew to Natal from where there was a vast expanse of sea between it and Dakar in West Africa, a distance of 1900 miles (3,057 km). Amelia and Fred Noonan took off for Dakar on 7th June at 3.15 am. Despite Noonan's excellent ability as a navigator they landed in Senegal about 80 miles (129 km) from Dakar. Amelia admitted when they landed that it was she who had made the error. They later flew on to Dakar where the plane was refuelled and checked over.

From Dakar Amelia Earhart and Fred Noonan flew the 1900 miles (1,834 km) to Gao and then over central Africa to Fort Lamy in Chad. On landing they discovered that the plane's landing gear had become defective. When this was attended to they sped on to Khartoum, to Massawa, Assab and over Saudia Arabia and the Arabian Sea. All this time their flight was given wide coverage in the press. When they landed in Karachi the plane was checked out and instruments adjusted. Following a stop there of a couple of days they again took off for Calcutta on 17th June, a distance of 1390 miles (2,237 km). The plane performed very well and flew through numerous squalls before landing.

Burma was their next destination and they set off in monsoon type weather, making a refuelling stop at Akyab. The weather was so bad that they had to return to Akyab having been in the air for some time. Next day they left in appalling weather and only got as far as Rangoon. From there it was on to Singapore and Bandung in Java where the engines were checked out. They had developed some instrument problems which took some days to iron out. Port Darwin Australia was their next port of call and to break the long flight they first landed at Koepang on the island of Timor and eventually reached Port Darwin on 27th June.

The most daunting part of the flight was now looming on the horizon. Amelia and Fred Noonan had already flown about 22,000 miles (35,404 km) and by this time were feeling the strain. After two days in Port Darwin they flew to Lae in New Guinea, a distance of 1200 miles (1931 km). Lae was the stopover from where they would set out on the most dangerous and difficult part of the entire flight across the Pacific to the tiny Howland Island. It needed expert navigation to find it, especially in poor weather as the island was only 3/4 of a mile (1.2 km) in length and 1/2 mile (0.8 km) in width. A flight between Lae and Howland Island, a distance of 2,556 miles (4113 km) had never before been undertaken.

World interest was now really focused on this section of the flight. To assist the aviators, the US Navy's USS *Ontario* would be stationed midway between the two locations while the Coast Guard Cutter *Itasca* would lie off Howland Island acting as a rescue ship if needed and with communications. It was hoped to take off from Lae on 1st July but the Weather Bureau advised against it. Howland Island was relatively a speck in the centre of the vast northern Pacific Ocean and required exceptionally accurate navigation to pick it up. From what was later learned Fred Noonan was having some difficulties in setting his chronometers due to radio problems he was experiencing. Because of the tense military situation between America and Japan it was vital that Amelia avoid the islands north of the Equator such as the Mariana and Marshall Islands, the property of Japan which the Americans saw as posing a threat to the control of the Pacific.

Amelia and Fred Noonan took off from Lae at 10 am on 2nd July, the Lockheed Electra loaded to capacity with fuel to give them every chance should bad weather impede their progress. In this situation the take off would be tricky due to the fuel weight and length of runway. They lifted off with little to spare. The District Superintendent of Civil Aviation for New Guinea who saw the take off described it as "hair-raising".

Weather reports were relayed to the plane at 10.22, 11.22 and 12.22 but no acknowledgement was received. However a brief transmission was made by Earhart around 3 pm to the radio operator at Lae saying that the plane was flying at 10,000 feet but was about to reduce altitude because of thick clouds ahead. All this time the American ships *Ontario* and *Itasca* , the latter off Howland were closely monitoring every signal and there was also a naval ship, USS *Swan*, between Howland and Miami in case Amelia missed the island.

Amelia Earhart made a further contact at 5.25 pm and gave her position in Latitude and Longitude which put the plane about 785 miles (1,263 km) from Lae and almost on course for Howland Island. The speed was 150 knots and the height 7000 feet. That position indicated that they were experiencing headwinds which had been forecast. Given

Amelia Earhart with her Lockheed Electra, the plane she flew on her ill-fated flight.
Smithsonian Institution
Photo No. A45874

57

the accuracy of the plane's position, the Lockheed Electra should be over the USS *Ontario* around 10 pm. The ship waited for a signal as did the other two ships but none came. However, a few minutes before 10 pm one of the *Ontario's* ship's officers on watch thought he had heard the noise of an aircraft. It was cloudy then and nothing could be seen despite the ship's searchlight sweeping the sky. About this time too a radio operator at Nauru Island station picked up a transmission with Earhart's voice saying "A ship in sight ahead". Neither the *Itasca* or the *Ontario* heard that transmission.

By now all those waiting to hear from Amelia Earhart were becoming anxious. They hadn't made contact since 5.25 pm. It was now 2 am and weather reports had been regularly transmitted to the plane but not acknowledged. Then at 2.45 am Amelia Earhart's voice broke through strong static on the *Itasca's* radio but all that could be deciphered was "cloudy weather ... cloudy". At 3 am the *Itasca* sent out the weather report giving clear skies, unlimited ceiling, calm seas with only 8 knots of wind. Amelia's voice was again picked up at 3.45 am saying "Itasca from Earhart, Itasca broadcast on 3105 kilocycles on hour and half hour". She repeated this request and mentioned "overcast". The *Itasca* did as she requested and asked her to transmit to them on 500 kilocycles in order that the ship's direction finder could get a fix on her. It was a futile request by the *Itasca* because Amelia's plane was unable to broadcast on that frequency.

As time went by the safety of Amelia Earhart and Fred Noonan was becoming top priority. During the next three hours nothing was heard from them but at 4.53 am Amelia's voice was picked up but her message could not be understood. She had now been in the air for about 18 hours and the sands of time were beginning to run out. She would have about four hours fuel left. She came through again at 5.12 am, this time in a clear voice and said "Want bearing on 3105 kilocycles on hour, will whistle in microphone". The *Itasca* was having difficulty in getting a bearing on the plane as Amelia's whistles didn't last long enough. Her voice broke in again three minutes later saying "About 200 miles out".

The *Itasca* was not having any luck in getting a bearing on the plane and at 6.45 am Amelia Earhart came on the air again, this time with a sense of urgency in her voice saying "Please take a bearing on us and report in a half an hour. I will make noise in microphone. About 100 miles out." She again whistled into the microphone but they could not get a bearing. The atmosphere now in the *Itasca's* radio room was extremely tense as the operator continued to send messages to the plane but without acknowledgement. Then Amelia came through again around 7.30 am, her voice displaying greater urgency and concern. She said "We must be on you but cannot see you but gas is running low. Have been unable to reach you by radio. We are flying at 1000 feet."

The situation was now critical but the *Itasca* was making no headway in getting a bearing on the plane. It continued to request Amelia to transmit on 500 kilocycles but she was unable to do that. At 7.57 am in a very clear voice Amelia Earhart came through on 3105 kilocycles saying "We are circling but cannot see island. Cannot hear you. Go ahead on 7500 kilocycles on long count either now or on schedule time of half-hour." The *Itasca* immediately did as requested and sent out the letter "A" continuously. Amelia replied at 8.03 am "We received your signal but unable to get minimum. Please take bearing on us and answer on 3105 kilocycles". The direction finder on Howland couldn't get a bearing. There was a total sense of frustration everywhere.

At 8.43 am Amelia Earhart called again in a voice that now had more than a sense of panic in it. "We are on the line of position 157-337. Will repeat this message on 6210 kilocycles. We are now running north and south". It was the last time that her voice was heard. There was nothing but static when the operator on the *Itasca* tuned in to 6210 kilocycles. He continued calling the plane but there was no reply. As time ticked away the *Itasca* crew knew that the plane's fuel couldn't last much longer. They also had the problem of interpreting exactly what Amelia Earhart meant by 157-337. There was no reference point given along this south east - north west line and therefore the plane could be anywhere along it.

Around 10 am on that fateful morning of 2nd July it was accepted aboard the *Itasca* that more than likely Amelia Earhart and Fred Noonan had come down in the sea. The ship's captain Warner Thompson advised Honolulu that he was about to begin a search for the missing aviators. He decided to search first in the area around Howland Island and then north west of it as indicated by the 337 reference. There had to be some starting point.

By now too the American Government had been fully alerted. They had co-operated with the round-the-world flight and as will later be seen probably had more than a passing interest in its success. Also of course Amelia Earhart and her husband George Putnam were good friends of President and Mrs. Roosevelt. The President on learning of the situation ordered the dispatch of the battleship USS *Colorado* , then on a training cruise off Hawaii to proceed to Howland Island and assist in the search. The *Colorado* had three catapult observation planes on board. The aircraft carrier *Lexington* and three destroyers were sent from the American west coast to the Howland Island area. The *Lexington* would take 10 days to get to the location.

There were a number of possibilities as to where the plane may have gone down, the Howland Island area being one. There were the Gilbert Islands too where they may have drifted to and got the plane down safely. The fear was that Amelia Earhart and her navigator may have been well off course north west of Howland Island and was obliged to land on one of the Marshall or Caroline Islands or ditch near one of them and then picked up by the Japanese.

As the *Itasca* began its search a long-range reconnaissance plane left Pearl Harbour to assist in the search. Later it encountered very bad weather as it flew a few hundred miles north of Howland between 2000 and 12,000 feet, weather that probably Amelia Earhart also experienced. The world press now carried the banner headlines of Amelia Earhart's disappearance. Many could not believe it. To them she was invincible. They expressed hopes that she and Noonan would be found. Hopes faded as day followed day.

As in all such cases there were many reports of signals being picked up, some no doubt being just hoax calls. Two amateur radio operators in Los Angeles said they picked up two SOS calls which gave the call sign of Amelia Earhart's plane KHAQQ. On 4th July a message was picked up by operators at the Warlupe Radio Station in Honolulu "281 ... north ...Howland ...KHAQQ ...beyond north ...don't hold with us much longer ... above ... shut off". An amateur radio operator in Oakland, California picked up a similar message. The *Itasca* was asked to search the area referred to but found absolutely nothing. It was accepted that hoaxers initiated the calls that were picked up.

George Putnam on hearing of his wife's disappearance could not believe that such a thing could have happened. He visited the famous flyer Jacqueline Cochran, a friend of Amelia who claimed to have extrasensory perception. She told him where she thought his wife had come down but when the area was searched nothing was found.

One of the most exhaustive searches ever up to that time had been made by ships of the US fleet including that by the carrier *Lexington* with 60 aircraft on board but no trace of Earhart or Noonan had been found or of any plane wreckage. By 18th July and area of 151,000 square miles had been searched and it was felt that to continue was impractical.

It was then that claims, counter claims, rumours and denials concerning the disappearance of Amelia Earhart took root in the world press. The hoaxes also increased as did the speculation that the plane had come down on one of the Japanese held islands and that Amelia and

Fred Noonan were held captive there. It was a reasonable speculation and one which to this day carries a certain amount of credence.

If one looks briefly at the political situation at that time in the Pacific especially where America and Japan were concerned it may be easier to consider that at least there was that possibility. The US had been very concerned about the build up of Japanese forces and it was felt that war was inevitable if control of the Pacific was to be kept in US hands. Following the First World War, Japan was mandated the Caroline, Marshall and Mariana group of islands, which formerly had been under German control. Although Japan was given control of the islands by the League of Nations they were expressly forbidden to use them at any time for military purposes. This was something that they disregarded. America was fully cogniscant of the strategic value of the islands to Japan in the event of war. Howland Island to the south east of the Marshall Islands was owned by the US and would be an ideal air base if such existed from which to attack the Japanese held islands if necessary.

It would appear that Amelia Earhart's round-the-world flight may have been used as an excuse for the US to put landing facilities on Howland Island. She wished to refuel there on the long leg of her flight from Lae in New Guinea. Whether or not she was asked by President Roosevelt to do this is not clear but at any rate he directed that work go ahead early in 1937 for a landing base to be built to accommodate the Lockheed Electra when it landed there to refuel. It was a golden opportunity for the US to build the air strip on the island for one of the greatest of its civilian pilots. Japan no doubt was not oblivious as to what may have been the real thinking behind it and consequently antagonism towards the US was exacerbated.

As mentioned at the outset of this story, the reticence of the American Government regarding their involvement with the Earhart flight led many to believe that it was something more than just another aviator anxious to achieve something that had never been done before. The press became extremely vocal when it was finally accepted that Amelia Earhart and Fred Noonan had gone forever. The *Oakland Tribune* began a series of articles in May 1938 on the disappearance. A journalist who managed to examine some Coast Guard files blamed the tragedy on the US Navy direction finder on Howland Island which when put aboard the *Itasca* was supplied with the wrong type of batteries resulting in the unsatisfactory operation of the equipment. The remainder of the series of articles was censored. A newspaper in Sydney, Australia called *Smith's Weekly* alleged that the US used Amelia Earhart's disappearance as an excuse to fly over the Japanese held islands and gather as much military information as possible. That paper also claimed that the Australian defence establishment had been informed of this.

In 1944 when the US invaded the Marshall Islands, Vice Admiral Edgar A. Cruise was informed by a native interpreter on Majuro Atoll that two fliers, a man and a woman had been brought to the Marshall Islands in 1937. This story was corroborated by another native of Majuro. The information was added to also in 1944 with the invasion of Saipan by the US when again it was stated that two aviators were brought to Saipan in 1937 where they were held. The woman, it was said, died from dysentery and the man was afterwards executed as being an American spy.

There was a rather convincing story told by a woman from Saipan named Josephine Blanco who later married and lived in California. She had been working for an American dentist in Saipan in 1946 and told him that she remembered in 1937 that two Americans crashed their plane. She was eleven years old then and remembered seeing a man and a woman, the latter was wearing a man's clothes and had closely cropped hair. They had been brought ashore and taken away by Japanese soldiers to the military barracks from where she heard shots.

An intensive investigation into Amelia Earhart's disappearance was carried out in 1960 by the Columbia Broadcasting System. The islands already referred to were visited and hundreds of people were interviewed. About thirty confirmed the story that two American flyers, a man and woman, had been brought to the island of Saipan where they died.

Two US Marines Bill G. Burks and Everett Henson Jnr. stated in 1964 that in July 1944 they belonged to a group who, in Saipan, recovered the remains of Amelia Earhart and Fred Noonan from an unmarked grave. The remains were placed in metal cans and brought to the US. This was never verified officially.

Over the years numerous stories have come to light regarding the mysterious disappearance of Amelia Earhart. Some no doubt were hoaxes. Others could have been planted by American Intelligence for general consumption to divert attention from getting any nearer the truth. There is little doubt but that the US and its military agencies were very interested in Earhart's flight and although her aeroplane had no known photographic equipment fitted, she and Fred Noonan were no doubt asked to keep their eyes and ears open especially when they flew across the Pacific. If American Intelligence had no involvement whatever, why all the coyness and reticence with regard to information on the flight, especially as the disappearance of the plane and its crew resulted in one of the greatest search operations up to then being undertaken by the US.

That the US had more than passing interest in Amelia Earhart's flight became very clear from 1968 onwards when the Freedom of Information Act became law. Since then numerous files on the flight were released by several government departments including that of Defence. Much has come to light but what finally happened the Lockheed Electra and its crew has never been disclosed. The US and Japanese Governments have not made a final pronouncement and so the mystery goes on. Numerous books, articles and film accounts have devoted great attention to what was the ultimate end to one of the greatest women pilots in the world but we still are no nearer a solution.

It is conceivable that the plane went down somewhere off Howland Island and sank without trace but it is more acceptable that the plane went well off course due to navigational problems on board, accentuated by poor weather conditions, and crashed near one of the Japanese held islands north west of Howland. The crew were more than likely picked up and held by the Japanese. Their fate is unknown. Unless irrefutable evidence from the US or Japan comes to hand or the wreck of the Lockheed Electra is found, the disappearance of Amelia Earhart and Fred Noonan will always remain a mystery.

SOPHIE PIERCE
(LADY MARY HEATH)

There are few whose triumphs over adversity exceeded those which epitomised the life and times of Irishwoman, Sophie Catherine Pierce, a late starter to aviation and better known as Lady Mary Heath. Few of her generation had the versatility and adaptability to live a life to the full that was often beset by sorrow and tragedy.

Her father Jackie could be described as a "wild man" and in an effort to give him some form of stability, his doctor father got him a job in the Kilrush, Co. Clare Branch of the Provincial Bank. Following a chapter of escapades he was removed from the bank and eventually inherited Knockaderry House from his maternal grandfather.

There he lived with the aid of a housekeeper named Kate Teresa Doolan whom he eventually married in the Dublin Registry Office on 29th May 1895. Sophie Catherine Pierce was born on 10th November 1896 and shortly after her birth, her mother was found dead in Knockaderry House. Jackie Pierce was picked up and charged with murder. He was found to be insane at the time and was confined to a mental institution for the rest of his life. Dr. George Pierce took responsibility for the infant Sophie and brought her to his home in Newcastle West, Co. Limerick where she was brought up by aunts.

When Sophie Pierce was of an age, she was sent to school in Dublin and later she enrolled in the Dublin College of Science where she obtained a degree in the early 1920s. Romance soon entered her life when she met a British Army officer named Elliott Lynn who was then home on leave from East Africa and staying with friends in Ireland. It was a whirlwind courtship, and after the marriage the couple went out to East Africa. Sophie was blissful happy at first and amused herself by writing poetry for the local Press. The marriage however didn't last and the couple parted.

Sophie returned home and then went to England where she published her poems. Following her marriage break up she bounced back and got very involved with athletics. She had a natural talent and on 6th August 1923 she set a world record for the women's high jump and also became British Javelin Champion in that year. She later became Vice-President of the Women's Amateur Athletic Association and played a major part in furthering the role of women in athletics.

It was during this time that she became interested in flying. She was returning from an Olympic Council Meeting in Prague in May 1925

*Sophie Pierce (Lady Mary Heath) at the height of her flying career.
Courtesy John Cussen, Solr. NewcastleWest.*

when she got into conversation on the plane with an RAF man named Reid. Having expressed an interest in aviation, he promised to help her learn to fly. She had a natural aptitude for flying and very quickly mastered her craft. She joined the London Light Aeroplane Club, becoming its first woman member and strongly encouraged more women to enter the world of flying. She had her first flight in August 1925 and quickly got involved with competitive flying.

Over the next few years Sophie Pierce broke many records. In 1926 she became the first woman to hold a British Air Ministry's Commercial Pilot's Licence or "B" Licence as it was then called. She held the world altitude record for light aeroplanes and was the first woman to make a parachute jump in April 1926. Meanwhile she married for the second time - to Sir James Heath, a wealthy English industrialist who was able to finance many of his wife's flying escapades. Sophie became Lady Mary Heath, a name she went under from then on, even though the marriage didn't last very long, law suits and notoriety bringing it to an end. Sir James commented at one stage "My wife has flown away in the clouds."

It was in 1928 that Lady Mary Heath gained international recognition when she made the first solo light aeroplane flight from Capetown to London in an Avro Avian III . She was later to create another altitude record for light aircraft by reaching a height of 23, 000 ft. She was constantly looking out for records to break, such as when she took off and landed in 50 different airfields and 17 'likely' landing fields in England in a single day with six refuelling stops. By now she had become a highly respected figure in aviation. She went to the US in 1928 and continued her success as an aviator there participating in flying demonstrations and giving lectures. While there she was seriously injured in an air crash in Cleveland in August 1929. Her plane crashed through a factory roof and although she seemed to make an excellent recovery and flew again, she was never quite the same person and appeared to suffer some personality change.

Even though she had two failed marriages, they did not deter her from entering into a third. She married a coloured aviator named Williams and they continued with flying exhibitions both in Ireland and in the US. For some time she was involved with Kildonan Aerodrome in North Co. Dublin and in the establishment of Dublin Air Ferries. She founded the Irish Junior National Aviation Club and her fame attracted many young would-be flyers to the airfield where she gave practical tuition in all aspects of flying during the Summer. In Winter she gave lectures on aviation in a hotel in Trinity Street, in Dublin.

Lady Mary Heath made a far greater contribution to aviation than she has been credited with. She was not an aviator just because she liked flying and used it for her own personal satisfaction. She passed on her enthusiasm to all with whom she came into contact and her great encouragement to women to take up flying was one of her many fine traits.

As the 1930s drew to their end, Lady Mary's health deteriorated. She became increasingly unstable and developed a drink problem which resulted in her last years being rather sad. In May 1939 she tripped and fell down the steps of a bus in London and died later as a result. She was only 43 years old.

She began life with a mentally deranged father who murdered her mother and now as she herself lay dying she decided to leave the world with a flourish. She arranged that her remains be cremated and the ashes brought from England by aeroplane to Newcastle West. The plane was to arrive at 12 noon and the ashes were to be scattered out immediately over the ground adjoining the houses in the south western angle of the Square. Sophie (Lady Mary Heath) had her reasons for having this done.

At that time in 1939 the south western angle of the Square was made up of (a) The house where she lived for many years; (b) the entrance to the house of Captain Richbel Curling (Agent of the Earl of Devon's Limerick lands) against whom she had a deep hatred; (c) the Church of Ireland, enclosed by a stone wall facing on to the Square.

It was the Captain's habit to walk out precisely at 12 noon each day, put his back against the Church wall and survey the scene. True to form the aeroplane passed low overhead and Curling couldn't avoid inhaling some of the ashes which kept him coughing for some time.

Sophie Pierce had made her final gesture from the air!.

THE BICYCLE MECHANICS WHO MADE AVIATION HISTORY

Although flying in lighter than air carriers such as balloons had been successfully carried out since the 1780s, it wasn't until 17th December 1903 that the first heavier-than-air aeroplane flight took place. Credit for this goes to two unassuming and rather shy Americans, Wilbur and Orville Wright who had little formal education and certainly no scientific training apart from what they taught themselves.

Their father Milton Wright was a member of the United Brethren and editor of its official organ, the *Religious Telescope*. Later he became a bishop in that sect. Little is known of their mother Susan Catherine Koerner Wright who, it was said, had a college education. There were five children in all, four boys and a girl named Catherine. They were a highly respectable middle class family but with little money to spare even when the father was made bishop. The children were strictly brought up, had very high standards of integrity and were taught to be careful with money, traits which the famous brothers never deviated from in later life.

Wilbur Wright was born near Richmond, Indiana in 1867 and Orville was born four years later. Following Orville's birth the family lived continuously in Dayton, Ohio. Although the other three children were sent to college, Wilbur and Orville left school at an early age, probably due to financial reasons. William left around the age of 14 while Orville attended high School for some time. Their father had a deep influence on their lives. A man of very strict principles, he was often attacked in the press for some religious decision he had taken. The brothers saw this and as a result had very little regard for newspaper comments. Throughout their lives they kept a very strained relationship with the press, often with hostility.

While still at High School, Orville Wright, aided by his brother Wilbur built a printing press. They wrote, printed, published and distributed several weeklies from 1889 onwards with titles such as *The West Side News, The Tatler* etc. The venture was reasonably successful but eventually Wilbur was forced to discontinue due to ill health.

In 1892 when the great cycle craze was beginning to hit America, the brothers Orville and Wilbur opened a small shop where they repaired and sold bicycles. This proved very successful and they expanded by manufacturing bicycles under their own trade names "Wright Flyer" and "Van Cleve".

The brothers gained valuable mechanical experience from this venture which certainly later stood them in good stead in their aviation work. As young men, their first contact with anything in the nature of aeronautics was probably in the flying of kites. However, their first real interest in flying came from Wilbur who had been following in the newspapers the exploits and experiments of the famous glider expert Otto Lilienthal. One day Wilbur was shocked to read that the German had been killed in a gliding accident. Orville now began to take an interest and over the next three years the brothers read all the important material on aeronautics that they could lay their hands on including a recently published book by a highly respected aeronaut and engineer named Octave Chanute. Some years later the brothers got to know him extremely well. He had been born in Paris in 1832 and brought to America by his parents in 1838 where he grew up. He had a very keen interest in gliding.

As the old century was drawing to a close, Wilbur and Orville Wright were taking the subject of aeronautics very seriously. There were numerous experiments being carried out throughout the world in an effort to solve the difficulties in flying a heavier-than-air machine. The Wright Brothers were very practical men and did not go along with many of the flying procedures being attempted. Wilbur was later to say "We were astonished to learn what an immense amount of time and money had been expended in futile attempts to solve the problem of flight."

They felt that continuous actual practice in the art of flying was a pre-requisite to any ultimate success. This called for courage and the brothers did not under-estimate the problems they faced. The year 1900 marked a milestone in the foundation work they had been giving their minds to regarding the practicalities of flying. It was now becoming a very serious business for them. They felt that they could and would solve the problems of getting a heavier-than-air machine to fly and were hooked by the prospect. They thought extremely highly of glider expert Chanute and entered into correspondence with him. He was then living in Chicago. Chanute was very impressed by the very professional and scientific approach they had shown in their very first letter.

One aspect of flying that the brothers had latched on to regarding the feasibility of flying a heavier-than-air machine was that of equilibrium. It was, Wilbur wrote "a stumbling block, this problem of equilibrium in reality constituted the problem of flight itself." They had confidence in their ability to solve this and were by now in the race to be the first to do so. When the busy season for bicycles had ended, the brothers decided to spend a few months away from Dayton to do some experimental work. They had built a bi-plane glider. This they now dismantled, packed carefully and set off with a tent on the very long journey to Kitty Hawk on the North Carolina coast where they were assured by the American Weather Bureau they would get strong and constant winds. They intended first to fly the glider as a kite with one of them on board. They later wrote that "suitable winds not being plentiful, we found it necessary, in order to test the new balancing system, to fly the machine as a kite without a man on board operating the levers through cords from the ground."

Before leaving Kitty Hawk, the brothers did some actual gliding with their machine, one of them lying flat on the lower plane, a successful move as it considerably cut down wind resistance. All in all the experiments were very satisfactory. They wrote and told Chanute of their experiments but were totally opposed to his giving details in aeronautical magazines to which he contributed articles. By the end of 1900 the brothers were very happy with their investigations and in the meantime they had built up a good correspondence with Chanute who was being extremely helpful. They had yet to meet the great man.

The Wright Brothers were very busy with their bicycle business, working very long hours and of course it was the finance from it that enabled them to indulge themselves in their aeronautical experiments. Aviation was beginning to take over their lives. Wilbur had a short article published in the *Aeronautical Journal* of London. The brothers had built a new glider on the style of their previous one but it had almost twice the surface area. In 1901, having made arrangements for running of the bicycle business in their absence they set out for Kitty Hawk after 4th July. They built a shed to shelter their glider and also lived there in rather austere conditions. They wrote and invited Chanute to come and meet them . He accepted their invitation and suggested that he may be able to help them by bringing along two glider experts. They reluctantly accepted.

Now with five people involved in the tests their hopes were high but things did not work out as expected. Having flown their glider first as a kite, they were disappointed with its performance and even with some alterations made they were not at all happy. Poor weather also hindered the work and they left Kitty Hawk very disillusioned on 22nd August, Wilbur later saying that no doubt man would sometime fly "but it would not be within our lifetime."

Moral was now very low and the brothers also felt that aeronautics was taking them away too much from their real livelihood. They also could see no financial gain at the time from their work, something that was anathema to them as they had always worked to make money. Chanute understood their point of view but knew that they had so much to offer and encouraged them to keep on trying.

He induced Wilbur to give an address to the Western Society of Engineers of which he Chanute was president. This was no mean feat to get Wilbur to agree because the brothers were very private people and not at all seekers of publicity. In his address Wilbur described in some detail their experiments of the previous two years. He was very well received but more importantly the address had given the brothers renewed interest in aeronautics. They were later to write:

"We had taken up Aeronautics merely as a sport. We reluctantly entered the scientific side of it but we soon found the work was so fascinating that we were drawn into it deeper and deeper."

They began more scientific experiments and built a small wind tunnel. In it they tried out the behaviour of some two hundred model planes and Chanute was of great assistance to them with mathematical calculations which they sent him. Despite all this, despondency again intervened. Flying had taken a new turn which seemed to cut across all that they were working for. A rich young Brazilian named Senhor Don Alberto Santos-Dumont who had been studying and living in Paris for ten years had come to the fore in the aeronautical scene. Airships had been his interest and on 19th October 1901 he flew an airship from St. Cloud to Paris, around the Eifel Tower and back to St. Cloud again. It gained him a prize of 125,000 francs and world wide publicity. As far as the general public were concerned he had flown and steered his craft from one location to another. He gained more popularity when he gave 75,000 francs from his winnings to the poor of Paris and distributed the remainder among his own employees who helped him with his success.

The Wright Brothers could see little light at the end of their aeronautical tunnel at that time. Aeronautics was costing them too much time already. Accordingly, they set out to expand their bicycle business further and ceased their experiments temporarily. When Chanute heard of this he was disappointed and suggested he may be able to get them sponsorship. He offered to introduce them to Andrew Carnegie. They studied the offer which was indeed very tempting. However, their integrity came to the fore. They felt that they couldn't accept that kind of financial assistance lest they should not be able to devote sufficient time away from their bicycle business and they turned the offer down.

Although they had decided to discontinue their experiments, the aeronautical attraction was too strong to resist and by the end of the year they had changed direction again and began making plans for the 1902 season at Kitty Hawk. Despite pressures from their own business they worked very hard on their aeronautical calculations and experiments which they continued to send to Chanute for observation and verification. By now, Wilbur Wright's address to the Western Society of Engineers had become a major talking point in aviation circles. Chanute had circulated copies to many of his friends and as a result the Wrights' work was being referred to in many journals. The brothers were extremely nervous of attracting too much attention and although they trusted their friend Chanute implicitly, nevertheless they felt they had to be cautious. With Chanute's help they had produced pages of figures from their experiments but did not want anyone else at this stage to benefit from them. They were also conscious of a barrier between themselves and the aeronautical engineers and academics.

In 1902 much of their time was spent assisting their father with problems in the church. Matters were eventually sorted out and they set out on 25th August for Kitty Hawk. They had built a new glider with a number of modifications. They had added a steerable rudder while the two main wings had a total surface area of 305 square feet. There were several other technical modifications made as a result of their experiments, including a three-fold control mechanism. Again, as previously, they tested the new glider as a kite and got good results. They then carried out some careful glides at altitudes of around six feet and were delighted with its performance. They were very careful men, taking one step at a time and their main idea was to get as much flying practice as possible. Once they had mastered this, they felt that the higher altitudes could then be undertaken with safety. They were very happy with the control they exercised over the machine after making around fifty glides in September.

Their friend Chanute with a colleague named Herring came as arranged to Kitty Hawk in October to try out two gliders of their own. The trials were not very satisfactory and after some time they left. In the last ten days at Kitty Hawk before the Wright Brothers left for home, they achieved even better results than they had obtained in the previous month. They made over 200 glides, one lasting twenty seconds and covering a distance of 622 1/2 feet (190 m). They were delighted with such results and returned home to Dayton confident that they could now control their machine.

By this time there were numerous people in the race to be the first to

The Wright brothers Orville and Wilbur

fly a heavier-than-air machine and the brothers were aware of it. As before, there were those in Washington working with power models but the Wrights were on a completely different approach using the glider as a base from which to develop. It was gliding that had first intrigued them when Wilbur followed the work of Lilienthal in the newspapers until the latter was killed in a gliding accident. In many ways it was a contest between those with a scientific training and two bicycle mechanics who had gained all their aeronautical experience and knowledge the practical way. The brothers were by now becoming very hopeful of their success and Chanute advised them of the necessity of having their work to date patented.

Already they began making plans for 1903. It was their intention to build a much larger machine and be fully able to control it. If successful, the next step would be to add an engine. While the brothers were working hard in preparation for their 1903 trip to Kitty Hawk, their friend Chanute had completed a five months tour of Europe, spurring up interest in an aeronautical competition which was to be held at the 1904 St. Louis Fair. France at this time was the most advanced in the aeronautical field and Chanute concentrated on the possibility of having a heavier-than-air machine success by then. The aviator, Captain Ferber, had also been studying that possibility.

During the tour Chanute referred to the approach of himself and the Wright brothers using the principle of a glider as a base. This was a totally different approach to that of Ferber and many more. Not meaning any harm, Chanute promised that when he got home he would send some rough sketches of his own and the Wrights' gliders to those interested. When they got them in France, they were published in an aeronautical magazine and also appeared in magazines in several other European countries and in London. The Wrights were not at all pleased and felt that Chanute had been too indiscreet and so they became more secretive.

The brothers were now coming under the spotlight and began to get many enquiries regarding their work to date. The Frenchman Captain Ferber wrote offering to buy their 1902 glider. They would not sell it and told them that they required it for practice in some of their 1903 experiments. They offered to build him one if he so wished in 1904. They had several requests for photographs but were very reluctant to face the camera.

One of their main concerns now was to get a suitable engine for their design. They wrote for details of engines to many manufacturers but none matched the specifications they required. They decided that the only way out was to build an engine themselves. This they did, assisted by their mechanic Charlie Taylor. Probably one of the great advantages the Wright brothers had over their rivals was that they were able to fabricate most of their requirements, being highly skilled mechanics. Also it kept their costs down to the minimum. Orville Wright designed the engine that was hoped would bring them success. It was 12 hp based in part on a Pope-Toledo car engine. It was fitted to the lower wing. Then there was the question of propellers. They were fully aware of their use in marine engineering but their design and performance in the field of flight was another matter. They succeeded in this also and were later to write about it:

"What at first seemed a simple problem became more complex the longer we studied it. With the machine moving forward, the air flying backward, the propellers turning sidewise and nothing standing still, it seemed impossible to find a starting-point from which to trace the various simultaneous reactions."

The new bi-plane had a 40 ft. wing span and was much heavier than the previous model. It had three main changes from the 1902 machine. There was an engine, propellers and sled runners - to prevent the machine from rolling over in a forward direction after landing. The engine drove two pusher propellers via bicycle chains. Their experiments through the winter of 1902 and early months of 1903 had given them great confidence of being close to success.

There were others in the race in America, among them Samuel Pierpont Langley whose work in the aeronautical field had been very well known. Recently he had received a grant of $50,000 from the American Board of Ordnance and Fortification to develop his heavier-than-air machine and was forced to remain tight-lipped about his progress which apparently hadn't been very inspiring.

Meanwhile Wilbur Wright was invited to present his second paper to the Western Society of Engineers in June 1903. He was not very happy doing this and it was actually to be the last talk he would ever give them. The subject matter concerned their experiments at Kitty Hawk in 1902. They were careful not to disclose too much data that might be advantageous to their competitors. The address was very well received and copies of it with photographs of the 1902 glider were circulated by Chanute to aeronautical people in Europe. The closer the Wright brothers were drawing near final success the more secretive they became. Mr. Chanute realised this when he asked for more details but was politely refused.

On 25th September 1903 Wilbur and Orville Wright arrived at Kitty

Hawk. Their intention was to be back in Dayton with the family for Christmas. Time was important for their flight trials but a considerable delay was encountered in assembling their new machine as great care had to be taken with every aspect of the work. Any speeding up could easily create a flying problem resulting in disaster. As it was, they ran into considerable difficulties before ever attempting an actual flight. In the course of pre-flight tests the propeller shaft broke off and the propellers were damaged. This necessitated Orville returning to Dayton, a distance of 800 miles to have replacement parts fabricated. He didn't get back to Kitty Hawk until 11th December.

In the intervening period there was some news in the press to the advantage of the brothers. The long awaited flight trial by Langley on 7th October was a failure as was also that on 8th December. The field now appeared to be clear. On 14th December they were ready to undertake a trial of their new machine. It stayed off the ground for just 3 1/2 seconds having travelled 100 feet (30 m). The result wasn't too promising and in the course of that brief trial two of the plane's struts were damaged but they repaired them.

The weather had now become rather inclement but they were determined to go ahead. Displaying immense courage under such conditions they decided that they would attempt another flight. On the morning of 17th December 1903 the Wright brothers felt that the time was right. There were five other people present on that momentous occasion, three from the Kill Devil Life Saving Station and two friends. This tiny group was to witness the first ever successful flight of a heavier-than-air machine called *Flyer*. It was a 12 second flight undertaken by Orville Wright, a flight that was to propel the brothers into the pages of aviation history. They had won. They flew successfully three more times, the last being the longest and flown by Wilbur.

It would be impossible for an outsider to express adequately the elation felt by the brothers who, against all the odds, unravelled the mystery of flying which man had been trying to do for centuries. Yet these two very modest men were very low key in their description of that unique moment of history,

"The first flight lasted only twelve seconds, a flight very modest when compared with that of birds, but it was, nevertheless, the first in the history of the world in which a machine carrying a man had raised itself by its own power into the air in free flight, had sailed forward on a level course without reduction of speed and had finally landed without being wrecked."

Due to heavy winds, the fourth flight ended with a rough landing

The first heavier-than-air flight by the Wright brothers. Orville is the pilot of the Wright Flyer while Wilbur watches history being made.

Courtesy National Air & Space Museum Washington DC.

Ref. (A267867B)

and caused slight damage to the now famous bi-plane, the product of the Wright brothers' genius. Unfortunately there was worse to come on that record-breaking morning. While they were standing around in a little group discussing the famous flight, the machine was hit by a freak gust of wind and damaged to the extent of not being able to fly again. How serious the damage was has never been defined because conflicting accounts of the incident came from the brothers. In their account of what happened and which appeared in *Century* Magazine in 1908 they stated that "the damage to the machine caused the discontinuance of the experiments" but sixteen years later Orville Wright wrote in *The Slipstream* "We estimated that the machine could be put in condition for flight again in a day or two."

Regardless of this the Wright brothers achieved what they had set out to do and sent a telegram home to Dayton telling of their success and requested that the press be informed accordingly. The local papers were first to get the news and eventually it filtered to the great nationals not alone around America but throughout the world. As always with such breath-taking news the reports contained many exaggerations and inaccuracies. Flight distances of up to three miles were even referred to.

Mrs. Wright notified Octave Chanute who was delighted with the news. He didn't have any details but was inundated with requests for them from aeronauts around the world. From a scientific point of view he was very anxious that august bodies such as the American Association for the Advancement of Science be given the details as to how the flight was achieved and in a letter to the brothers he urged them that they present the facts in an address to the Association. It was a logical request coming from a man of science but the Wright brothers were not scientists and apart from their close relationship with Mr. Chanute they favoured giving scientists and academics a wide berth at least for the present. Also while they themselves had become the 'first to fly', such a scientific achievement was to them, to a great extent, a secondary matter relative to the finance they hoped to gain from it. Accordingly they turned down the request to address the American Association and also refused to give any technical details or photographs of their machine.

It was a big disappointment to the scientists but one which Chanute probably understood. The inaccuracies which kept appearing in the press regarding their first flight irked the brothers and they decided to put the matter to right once and for all. They sent a full account to the American press and sent a similar document to the Royal Aeronautical Society in London and to the French Aeronautical magazine *L'Aerophile*.

*Photo of one of the six Wright Flyers manufactured by short Bros. Belfast for the Wright brothers in 1909.
Courtesy Short Bros. plc. Belfast*

They very kindly sent letters also to various interested aeronauts including Frenchman Capt. Ferber.

That text of the Wright brothers' achievements on 17th December at Kitty Hawk was circulated and printed in full in many papers. Due to the importance of its contents, it is given here taken from the *Chicago Daily News* of 6th January 1904.

(By the Associated Press)

Dayton,O., Jan.6 - Wright brothers, inventors of the flying machine which has attracted such widespread attention recently, today gave out the following statement, which they say is the first correct account of the two successful trials:

"On the morning of Dec. 17, between 10:30 and noon, four flights were made, two by Orville Wright and two by Wilbur Wright. The starts were all made from a point on the level and about 200 feet west of our camp, which is situated a quarter of a mile north of the Kill Devil sandhill in Dare County, North Carolina.

"The wind at the time of the flights had a velocity of 27 miles an hour at 10 o'clock and 24 miles an hour at noon, as recorded by the anemometer at the Kitty Hawk weather bureau station. This anemometer is 30 feet from the ground. Our own measurements, made with a hand anemometer at a height of 4 feet from the ground, showed a velocity of about 22 miles when the first flight was made and 22 1/2 when the last flight was made. The flight was made directly against the wind. Each time the machine started from the level ground from its own power, with no assistance from gravity or other source whatever.

"After a run of about 40 feet along a mono-rail track, which held the machine 8 inches from the ground, it rose from the track and under the direction of the operator, climbed upwards on an inclined course, till a height 8 or 10 feet from the ground was reached, after which the course was kept as near horizontal as the wind gusts and the limited skill of the operator would permit.

"Into the teeth of a December gale the 'flyer' made its way, with a speed of 30 to 35 miles an hour through the air. It had previously been decided that for reasons of personal safety these first trials should be made as close to the ground as possible. The height chosen was scarcely sufficient for maneuvering in so gusty a wind, and with no previous acquaintance with the conduct of the machine and its controlling mechanisms. Consequently the first flight was short.

"Successful flights rapidly increased in length and at the fourth

Photo of page from the Order Book of Short Bros. in 1909 when the Wright brothers placed an order for 6 Flyer aeroplanes.

Courtesy Short Bros. Plc. Belfast

trial a flight of 59 seconds was made in which the machine flew a little more than half a mile through the air and a distance of more than 852 feet over the ground. The landing was due to a slight error of judgement on the part of the navigator. After passing over a little hummock of sand in an attempt to bring the machine down to the desired height the operator turned the rudder too far and the machine turned downward more quickly than had been expected. The reverse movement of the rudder was a fraction of a second too late to prevent the machine from touching the ground and thus ending the flight. The whole occurrence occupied little if any more than one second of time.

"Only those who are acquainted with practical aeronautics can appreciate the difficulties of attempting the first trials of a flying machine in a 25-mile gale. As winter was already set in, we should have postponed our trials to a more favourable season but for the fact that we were determined before returning home to know whether the machine possessed sufficient power to fly, sufficient strength to withstand the shock of landings and sufficient capacity of control to make flight safe in boisterous winds as well as in calm air. When these points had been definitely established we at once packed our goods and returned home, knowing that the age of the flying machine had come at last.

"From the beginning we have employed entirely new principles of control, and as all the experiments have been conducted at our own expense, without any assistance from any individual or institution, we dot not feel ready at present to give out any pictures or detailed description of the machine."

One very important reason the Wright brothers resorted to secrecy regarding their machine and its technical details was the matter of patents. They had filed patents in many countries where it was thought their work might be illegally copied and had no confirmation as of yet regarding the patenting. They were extremely lucky too in those early years to have a man of the stature of Octave Chanute not alone advising them on such matters but who could also vouch for their achievements as their secretiveness often cast doubts over their achievements.

They added further confusion and contradiction to their policy of secrecy when out of the blue they invited the press to a demonstration flight of their new machine close to home in May 1904. To make matters worse, the brothers ran into several problems and the flight was a fiasco and never got off the ground. It was unfortunate and it increased the doubts for some time to come as to the credibility of their claims. Many changes were made in *Flyer II* and they undertook several successful flights later in the year, the most successful being on 9th November when the flying field was circled several times, covering a distance of over 3 miles in five minutes, four seconds. A further version of their biplane was flown by them on 5th October of the following year 1905 and Wilbur Wright flew for 38 minutes.

Although the Wright brothers had made it possible in 1903 to fly a heavier-than-air machine, they kept their know-how to themselves and it wasn't until Santos-Dumont in 1906 made data of his successful flights public that aviation really made strides. The brothers were obliged to abandon their bicycle manufacturing business in 1904 due to the great pressures they were under and devoted all their time to the business of aviation. They designed a new version of the *Wright Flyer* and in March 1909 placed an order for the manufacture of 6 of them with Short Brothers, Belfast. They had many further successes, including a Wright design purchased by the American military authorities which Orville had flown for 1 hr. 12 mins. 20 secs on 27th July 1909. It was called *Miss Colombia* and the brothers received $25,000 for it plus another $5,000 for exceeding the speed specification required. The brothers great achievements also brought them much commercial success abroad.

Wilbur and Orville Wright, two bicycle mechanics, had shown to the world on 17th December 1903 that with no great formal education, they became the first to solve the problems of getting a heavier-than-air machine into the air and flying it successfully. Sadly Wilbur died of typhoid fever in May 1912. His brother Orville lived another 36 years.

THE FIRST ATLANTIC CROSSING

When it is asked who made the first crossing of the Atlantic in a heavier-than-air machine, invariably the reply given is that it was Alcock & Brown. This is not correct. It was Lieut. Commander A.C. Read and his American Navy crew of five. They achieved this in a Navy-Curtiss flying boat, making the necessary refuelling stops on the way. Its significance can be viewed in the context of what Charles Lindbergh said about it years later, when speaking of his own Atlantic solo flight.

"I had" he said "a better chance of reaching Europe in the *Spirit of St. Louis* than the NC boats had of reaching the Azores. I had a more reliable type of engine, improved instruments and a continent instead of an island for a target. It was skill, determination and a hard-working crew that carried the NC-4 to the completion of the first trans-Atlantic flight."

To be first to make an air-crossing of the Atlantic, was not alone every aviator's dream, it was also the dream of nations. As early as 1910, serious thought was being given to it by the use of balloons and non-rigid airships. In 1913, Lord Northcliffe, who owned various publications including the Daily Mail, threw down a challenge in that paper to would-be aviators. He offered an award of £10,000 (then $50,000) to the first to cross the Atlantic by plane, either way between the North American Continent and any point in Ireland or the UK, within 72 consecutive hours. Aviators would be obliged to complete the flight in the same craft in which they began. Intermediate stoppages were allowed only on water.

The notice in the Daily Mail attracted enormous interest. American Glen Curtiss, the first man to take off in an aeroplane from water, was extremely interested . Highly respected, the US Navy gave him the assistance of his close friend, Lieut. John Towers, in the construction of a flying boat. It would be called *America* and would attempt the Atlantic crossing. It was assumed that both Curtiss and Towers would be the plane's pilots when it was ready to fly. Mrs. Curtiss had other ideas. She was not at all keen on her husband embarking on this flight and he accordingly withdrew. Towers was refused permission by the Navy to attempt the crossing but was allowed to act as adviser. One of two selected was famous English aviator John Cyril Porte, formerly of the Royal Navy. The *America* was completed, her flight trials satisfactory and 15th August 1914 was fixed for her Atlantic attempt. On 3rd

August, Germany declared war on France and on Britain the following day. It was the end of the Atlantic crossing plan for some years. The *America* was sold to Britain, to serve as a prototype for 50 seaplanes to be built by them.

Increased German submarine activity during the war was posing problems for both Britain and France. Patrol planes were shipped by the US as quickly as possible but this was not adequate. The Royal Navy was turning out small flying boats, under the supervision of Commander Porte but were very limited in their endurance and depth charge capacity. America, in its determination to assist the Allies, decided to build long range flying boats, capable of being armed with bombs, guns and depth charges.

They were to be known as Navy Curtiss planes, the first called NC-1 and so on. They had to be capable of crossing the Atlantic, probably from Newfoundland to Ireland. The original design was later altered to a flying boat that would be capable of crossing the Atlantic via the Azores.

The US Navy finally accepted the design and placed an order for four with the Glen Curtiss Company. Several manufactures supplied parts, such as wings, hulls and tails. They were shipped to the Curtiss plant at Garden City, Long Island. A large hangar was built nearby at the Naval Air Station, Rockaway Beach. Work progressed rapidly and on 4th October 1918, the NC-1 was flown for the first time by Commander Richardson and his crew. The war ended on 11th November and the long-range anti-submarine flying boats were no longer required.

However, the challenge still remained to be first to fly the Atlantic and the US Navy decided that, because of the experience already gained, it would meet that challenge. Lieut, Richard E. Byrd, a naval aviator engaged in the study of crashes, requested that he be detailed to fly the Atlantic in the NC-1 when it was ready. He was refused. Commander John Henry Towers, a naval aviator taught to fly by Glen Curtiss in 1911, was assigned to head the operation. Meanwhile, the NC-1 established a world record on 27th November 1918, by carrying 51 men into the air.

Following further flight checks, modifications were necessary to the NC-1. To the concern of many, they were being made at a leisurely pace. Other countries were aiming to be the first to make an Atlantic crossing, including Britain. In December 1918, Commander G.C. Westervelt of the US Navy returned from Europe to America. He was concerned at the Atlantic crossing attempts being planned. He submitted a report to the Trans-Atlantic Planning Committee of the US

Navy. They fully endorsed it. It was then given to Secretary Daniels of the Navy with the following addition:-

"As it seems probable that Great Britain will make every effort to attain the same relative standing in aerial strength as she has in naval strength, the prestige that she would attain by successfully carrying out the first trans-Atlantic flight would be of great assistance to her. ... In view of the fact that the first successful airplane was produced in this country and that the United States developed the first seaplane, it would seem most fitting that the first trans-Atlantic flight should be carried out upon the initiative of the United States Navy."

The basic plan was that four flying boats would attempt the Atlantic crossing from Newfoundland via the Azores. A flotilla of destroyers would be located along the route, fitted with special radio equipment, meteorological apparatus and star shells. The separation between destroyers would be 50 miles (80 km). Planning was meticulous and pressure was intense to have the four flying boats ready by a fixed date in May 1919. The ice would be broken up by then and the period of darkness not too great. There would also be a full moon on that date. The flying boats had a cruising speed of about 75 mph and it was arranged to arrive in the Azores in daylight. It was necessary to consult with the Allies. St. John's Newfoundland was a British Colony and it was from there that the flight would begin. Canadian and Portuguese ports would also be required to co-operate.

June or July would have been better months in which to make the attempt but time was of the essence. By now, there were nine British entries for the Daily Mail prize, the rules of which had been changed by Lord Northcliffe after the war. It was no longer permitted to land on water and also "machines of enemy origin" were excluded. This ruled out the NC flying boats and also the large German bombers.

It did not deter the Navy Curtiss Company, who were working all hours on modifications to the NC-1. A four-engined concept had worked well on the NC-2 and this was now adopted for the other planes. From the beginning, the programme was hit with problems. In March 1919, a storm dragged the NC-1 from her moorings, damaged the hull and shattered the lower port wing. When experiments with the NC-2 were completed, it was decided to transfer one of its wings to the NC-1 whose hull was being repaired. The attempted Atlantic crossing would now be made with three flying boats.

The 5th May was fixed for the first stage of the flight, that from Rockaway Beach to Halifax, Nova Scotia, a distance of 540 miles (86?

Lieut. Commander A.C. Read was US Naval Aviator No. 24
The National Archives, Atlanta, GA. USA
Photo No. 80-G-431548

Km). Commander Towers, who had overall control of the three flying boats, would travel on the "flagplane, "NC-3. Chief pilot of the NC-1 would be Patrick N.L. Bellinger. H.C. Richardson would fill a similar position on the NC-3, while Lt. Commander A.C. Read would be in charge of the NC-4. The flying boats had an endurance of 20 hours.

The US Navy decided on a backup in the form of a non-rigid airship called C-5. It was 192 feet (58.5m) in length and had two 125 hp. engines. Underneath was slung a 40 ft. (12.19m) control car with a crew of four. It had the capability of flying the Atlantic.

It was not until 30th April that the NC-4 was completed. When she was put on water she leaked badly. The hull and wing support struts were also damaged when she slipped on her beaching carriage. Furthermore, the control cables to the tail were carried away as she taxied in following her first flight. It was not a very auspicious beginning. Towers was happy with the NC-3 but now the NC-4 required considerable attention.

One could not be faulted for believing that there was a jinx on the entire American plan to cross the Atlantic. All was ready on the night of 4th May. During the early hours of the following morning, a fire broke out in the hanger which housed both the NC-1 and NC-4. Before it could be contained, sections of the tail of the NC-4 and an entire wing of the NC-1 were destroyed. The Navy-Curtiss men worked feverishly to offset the damage. Repairs were carried out and a new wing was fitted, taken from the NC-2.

Commander Towers told the press that the flight was postponed indefinitely. It wasn't the true story but it diverted their attention to the land planes. They were already in Newfoundland waiting for an improvement in the weather, before attempting the crossing to Ireland. On the morning of 8th May, with little or no press about, the three flying boats took off from Rockaway Beach at 10 am (Eastern Standard Time), having had a favourable weather report. They were bound for Halifax, Nova Scotia. The route chosen for the trans-Atlantic crossing attempt was Rockaway - Halifax-Trepassey Bay(Newfoundland) - Ponta Delgada (Azores) - Lisbon - Plymouth. The three flying boats under Commander Towers flew in a "V" formation. The "flagplane" had on its starboard, the NC-1 and on its port the NC-4.

Albert Cushing Read, central to this story had been born at Lyme, New Hampshire on 29th May 1887. Having graduated from the US Naval Academy in 1907, he was commissioned Ensign in September 1908. He began aviation training in July 1915, becoming Naval Aviator No. 24. When the US entered the First World War in 1917, he was

The NC-4 crew left to right: Chief Mechanic's Mate – Eugene S. Rhoads; Lieut. J.L. Breeze - Reserve Engineer; Lieut. Walter Hinton - Pilot; Lieut. Elmer Elmer Strone - Pilot; Lieut. Commander A.C. Read - Flight Commander. Photo No. 00029643 Smithsonian Institution.

promoted to the rank of Lieut. Commander and put in charge of the New York Militia Air Station at Bayshore, Long Island. Now, he and his five man crew were attempting to be part of a team that would hopefully make aviation history. The crew were very experienced. Lieuts. Elmer Stone and Walter Hinton were the two pilots. James Breeze, the engineer, was assisted by machinist's mate Eugene Rhoads and the Radio Operator was Ensign Herbert Rodd. The plane commander on each of the flying boats also acted as navigator.

Read in the NC-4 made an entry in his log at noon . "Passing Montbank Point. Sun came out". Ahead of him was the NC-3, while the NC-1 was some distance at the rear. Soon he began to experience problems and the other two planes forged well ahead of him. The centre pusher engine had to be switched off due to dropping oil pressure. Read decided to continue on to Halifax without it. Later on, a second engine gave trouble and the NC-4 was forced to come down and taxi to Chatham. Read and his men were bitterly disappointed. Towers heard of their predicament when he landed at Halifax. He later reached Trepassey Bay in Newfoundland as did the NC-1. They were forced to remain due to weather conditions. Had they been good, the two planes would have taken off for the Azores.

Because of the NC-4's troublesome record, she was dubbed "Lame Duck." Nevertheless, repairs went ahead quickly. Within two days one of the faulty engines had been replaced and the other repaired. Due to bad weather, it wasn't until 14th May at 9.16 am that the NC-4 took off. and reached Trespassey Bay on 15th May at 5.41 pm. The presence of three flying boats in the bay quickly attracted press attention. On the morning of the 15th it had become apparent that something was afoot, with the arrival of the airship C-5 at St. John's Newfoundland. It had been flown in secrecy from Montauk. When the *Chicago* sailed into St. John's to act as base ship for the airship, it confirmed the view that the C-5 was about to attempt an Atlantic crossing.

Now there came further trouble. A sudden storm arose and struck the location where the C-5 was moored. Before personnel could deflate her, she was torn from her moorings and blown out to sea, never to be found.

An improvement in the weather came on 16th May, and the three flying boats got ready to take of. The "flagplane", NC-3, was first to move along the water followed by the NC-4. As Read got airborne, he noticed that the NC-3 appeared to be having difficulty in taking off. The flying boat was actually overloaded and Towers was forced to off-load an emergency transmitter, a small wooden chair and a box of tools. To reduce the load still further, he had his chief flight engineer also taken off. Meanwhile, Lt. Commander Read returned to base and at 6 pm, the three eventually got away. The NC-4, with its earlier problems now solved, was in good shape and Read was in a confident mood. His plane was faster than the other two and quickly forged ahead from them, so much so that he circled and came behind them. As darkness came, it was difficult to see the other planes. He asked them to put on their lights but nothing happened. He decided to go it alone.

The destroyers based along the route were required to report the passing of the planes. Following each report, the next destroyer in line would commence firing its star shells. Read and his crew flew on through the night and at 6.30 am (Azores time) picked out destroyer No. 16. The weather deteriorated and No. 17 could not be seen. They continued on through thick fog and then the weather cleared somewhat. The NC-4 was at 3000 ft. but could not find any destroyer. At 9.27 am through an opening in the clouds Read saw land. When he descended to 200 ft., he discovered that they were over the southern tip of Flores about 250 miles (402 Km) from their destination. They sighted destroyer No. 22. As they came abeam Fayal Island the weather again deteriorated and Read decided to land at Horta. It was there the base ship *Columbia* was stationed. The NC-4 was brought down safely and the first leg of the trans-Atlantic crossing was successfully achieved in a flying time of 15 Hrs. 18 Mins.

All had not been well with the other two flying boats. The NC-3 was in difficulty. On take off, the plane's electrical circuitry got extremely wet. As a result, the lighting on the wing tips and cockpit was put out of action. When darkness fell Towers climbed to 4500 ft. to avail of the light of the moon. As dawn broke, the weather worsened and clouds began to thicken. He decreased altitude and through the haze picked out the outline of a destroyer. He took it to be No. 15 and altered course accordingly. He was incorrect.

By 11. am, having seen no further destroyers, Towers knew that he must be off course and in deep trouble. There was only two hours fuel remaining in the tanks. He decided to land and make some observations. Unfortunately, as the plane approached the water surface, he saw huge swells but it was too late to get the pilot to ascend again. As the plane hit the water, it bounced back into the air before landing. The force of the impact buckled some struts on the forward centre engine, control wires were strained and part of the hull frames split. There was absolutely no hope of being able to take off again. The NC-3 was still about 205 miles (330 Km) from Ponta Delgada. That evening, the weather deteriorated and it was the great seamanship of the crew that kept the plane afloat throughout the night. Destroyers searched without success and Towers was obliged to taxi to his destination. It was the end of the NC-3's attempt at the Atlantic crossing and a bitter disappointment.

What of the third flying boat NC-1 and her crew? They had set out with great hopes . Prospects looked good throughout the day and all through the night. They had picked out destroyer No. 17 while flying at 600 ft. Shortly afterwards they ran into fog and went up to approximately 3200 ft. The fog began to thicken further and in the interest of safety, descended again and landed on a comparatively rough sea.

It was an unenviable position to be in. The craft was tossed about in heavy seas. The sea anchor was lost shortly after landing but the crew were fortunate in their make-do substitute, a metal bucket which helped enormously. Due to the roughness of the sea, there was no hope of taking off again. The only thing to do was to sit tight and hope that NC-1 would be seen by some passing ship. Around 1800 GMT. on 18th May they were in luck when a steamer called *Iona* found them and took them in tow. That night an American destroyer took charge of the NC-1, which was now in poor shape. Lieut. Commander Bellinger and his crew were taken to Horta , disappointed by events but nevertheless glad that they, like the NC-3 crew, were safe.

Meanwhile, some land plane aviators ,who had also been attracted to Newfoundland as a potential departure point, had been observing the movements of the naval flying boats. In an attempt to beat the US Navy, two pilots, Harry Hawker and Commander MacKenzie Grieve, R.N. took off in a Sopwith bi-plane on 18th May. An hour later, William Morgan, a pilot who had lost a leg in the war and Frederick Raynham followed suit in a Martinsyde aircraft. The latter two crashed on take off and were injured. A week went by and nothing had been heard of Hawker and Grieve. Then, a message from a Danish tramp steamer, which had no radio on board was received by a semaphore station in the Hebrides, that they had picked up the crew of the Sopwith. They had ditched their plane when the aircraft engine gave trouble. The Daily Mail, in recognition of Hawker and Grieve's gallant effort, awarded them a consolation prize of £5,000.

A decision had now to be made by the US Naval Authorities if Commander Towers should join Read on the remainder of his flight. It was decided that Lieut. Commander Read would continue as the officer-in-charge. Towers was detailed to travel to Plymouth via destroyer.

The NC-4 at Lisbon 27th May 1919 following the first-ever crossing of the Atlantic.
The National Archives, Atlanta, GA. USA
Photo No. 80-G-650875

Having stayed in Horta from 17th to 20th May, the NC-4 flew to Ponta Delgada. This only took 1 3/4 hours and it was not until 27th May that the second stage of the crossing to Lisbon was undertaken. It proved to be an uneventful journey, with no difficulty in picking out the destroyers and marker buoys along the route. Nine hours and forty four minutes later, the NC-4 arrived in Lisbon to a great welcome.

The final goal was the crossing from Lisbon to Plymouth, which was begun on 30th May. They were only in the air for about an hour and a half, when a leak was discovered in a water jacket, forcing them to land for repairs at the mouth of the Mondego River. Some six hours later, NC-4 took off again for Ferrol Harbour where the crew rested until around 0630 the following morning. The tide was then favourable to set out for Plymouth . The weather, despite the time of year, was blustery but at last the English Channel was crossed. Lieut. Commander Read and his crew arrived to an ecstatic welcome at the Plymouth Sound around 1315 GMT. The flight distance from Rockaway, New York to Plymouth was 3925 miles (6316.5 Km) and they had covered it in 57 1/4 flying hours.

When the success of the airmen became known, messages of congratulations poured in from many parts of the world. America was extremely appreciative of their great achievement. President Woodrow Wilson in his cable said "We are all heartily proud of you. You have won and deserved the distinction of adding still further to the laurels of our country." Later the crew travelled to London, where they received a heroes' welcome from an enormous crowd, which included the Prince of Wales and Winston Churchill.

In a few weeks, their great achievement was overshadowed by Alcock and Brown, who successfully made the first non stop air crossing of the Atlantic and by the R-34 airship's flight from Scotland to New York and back to England. The flying boat men never got the recognition they so richly deserved. To this day, their great success is little known to the general public.

For his outstanding achievement and contribution to Naval Aviation, Lieut. Commander Read was awarded the following honours:- Distinguished Service Medal, Legion of Merit, NC-4 Medal (US); Tower and Sword (Portugal); RAF Cross and CBE (British). Following the famous flight, Albert C. Read took the NC-4 on a recruiting tour of 39 cities. He rose to the rank of Rear Admiral, serving as Commander, Fleet Air Norfolk, during the Second World War. He later retired and died on 10th October 1967 in Florida.

THE IRISH FLYING ACE

Whenever World War I Air Aces are referred to, names such as the German Baron Von Richtofen (The Red Baron) with his tally of 80 aircraft shot down or Rene Fonck the Frenchman with a score of 75 invariably come to mind. Yet, there was one who followed close on the heels of these men but who never attracted the same publicity. He was Irishman Major Edward "Mick" Mannock, VC; DSO (2 Bars); MC., Britain's top air ace in that war despite the disadvantage of being blind in one eye.

His parents had met while the Second Inniskilling Dragoons were based in Ballincollig, Cork. His mother was Julia O'Sullivan before her marriage to Corporal Mannock. They had five children, three of them girls, the eldest was Jesse. The two boys were Patrick and Edward, the latter being born at Preston Cavalry Barracks Brighton on 24th May 1887. Later the regiment moved to India where young Edward spent some of his formative years. There he developed an infection in his left eye. Despite medical attention he eventually lost the sight in it. As a young boy growing up he never displayed any tendency whatever that showed he would make a leader of men and an impassioned fighter pilot.

He was attracted to birds and animals and read a lot which was in sharp contrast to his father who had no interest in such things. In fact Corporal Mannock was somewhat of "a rough diamond" treating his family as if they too were in the army. However, Julia Mannock stood little nonsense from her husband where her children were concerned and always stood up for them especially Edward because of his disability. He liked music and taught himself to play on an old violin his father reluctantly bought him following pressure from his mother. With it he would often entertain his colleagues in the years ahead.

When the Boer War broke out, the Inniskillings were posted to South Africa and the families of the men were returned to England. Mrs. Mannock now had the responsibility of raising her children alone in Canterbury. When the war ended, her husband returned but couldn't settle down. He was a changed man who had become disillusioned with army life. He eventually left the service and also walked out on his wife and five children. She was never to see him again. Later her wayward husband contracted a bigamous marriage and raised another family.

Julia Mannock was a good mother who, many years later, was

described by the wife of her son Patrick as "a lovely homely lady but she frightened me with what she had to say about her husband who'd run off and left her with all the children to bring up. She once snatched up a broom and drove the handle straight through his picture."

When Edward's older brother Patrick had finished his schooling he got a job as a clerk with the National Telephone Company. Edward himself longed for the day when he too could leave as he found much in his schooling that he couldn't relate to. He was keen on English and reading but not on arithmetic. At the age of thirteen he left and got a job with a greengrocer. It was hard work but it helped his mother to keep food on the table for her family. Edward eventually left that job for a better one as a barber's assistant. The loss of one eye was a considerable handicap as he grew up but he gradually overcame it and even played cricket

Later, on the advice of his brother, he joined the local telephone company. His office duties with the phone company were not greatly to his liking. Pages and pages of figures made very little sense to him. He struggled on for some time and meanwhile joined the 2nd Home Counties Field Ambulance R.A.M.C. Territorial Unit. This appealed to him and nearly all his spare time was taken up with it. He also had a uniform and felt that he was now getting somewhere.

Not happy with office work in the telephone company he got himself transferred to outside work with the line crews. This meant working away from home and finding his own accommodation. In this regard he was exceptionally lucky as he got lodgings with a couple named Eyles in Wellingborough, Northampton. They treated him like a son and he was very happy with them.

By now Edward Mannock had gained considerable confidence and independence and the loss of the sight in his left eye was proving to be less of an obstacle. He had been in the Territorials for some time, eventually gaining the rank of Senior NCO. He played cricket rather well and joined a debating society. He was very much a Socialist. He had a very full life yet it wasn't enough and he longed for more excitement and an improvement in his situation. One day, to the great disappointment of the Eyleses, he told them that he wished to go abroad and improve himself. They very kindly helped him with the passage money which he promised to repay as soon as was possible. He set out for Istanbul on a tramp steamer and arrived on 1st February 1914 with fifty shillings in his pocket

His experience with the Telephone Company back home stood him in good stead because he shortly got a job in a supervisory capacity with the English Telephone Company who were laying phone lines for the Turkish Government under contract. The urgent work went on all through the Spring and Summer of 1914.

War clouds were by now looming and Mannock could sense this. Germany was becoming increasingly influential in Turkey and when the war did break out the Germans had a great say on economic and military matters. The Turkish papers became very pro German and the Turks became hostile to the British. The British Ambassador and his staff eventually left for home. The English Company which Mannock worked for refused to take orders from the Germans, who had by now, taken control of it and he with many of his colleagues imprisoned. They were poorly treated especially with regard to food. All this was certainly not what Edward Mannock had in mind when left England to seek his fortune.

Later the English were moved to an open camp which was surrounded by wire. Mannock crept out from there one night and sought out a Turkish friend Ali Hamid Bey who had worked with him in the Telephone Company. Bey gave him some badly needed food which he brought back to the women and those sick in the camp. On the next occasion he was caught and placed in solitary confinement, which affected his health. After some months the English were released through a prisoner exchange scheme arranged by the American Red Cross. It took them two months of very difficult travel to get back to England. By now Mannock was a very disenchanted man filled with an enormous hatred of the Germans. This hatred of the Hun was something that would often surface later. Following a brief period with his family and with his 'second family' the Eyleses he rejoined his old unit in the Royal Army Medical Corps. However, he sought an interview to put his case for transfer to the Engineers because he knew that if he had stayed with the RAMC he would sooner or later be sent to France to look after the wounded, many of whom would be German prisoners, "I don't want to nurse sick and wounded Huns, I want to fight them" he said to the officer interviewing him.

His request for a transfer was granted after a few months and he was posted to the Royal Engineers Signal Depot at Fenny Stratfold. He was commissioned in June 1915. By now his hatred of the Germans had deepened and he was determined to do everything possible to get into action against them and have them defeated.

Edward Mannock had a total dislike for the English class system and often let it be known. He was also critical of the way the war was being run and in the mess would speak out against it. This sometimes put him on a collision course with other officers who looked upon him as unstable. In any discussion he always gave as good as he got and did not hesitate to speak his mind.

It was while home on leave that the course of his life changed towards aviation following an accidental meeting with an old friend of his named Eric Tompkins who was in the Royal Flying Corps. Tompkins suggested to him that he should seek a transfer to the RFC if he wanted to get into action quickly. There were problems as Edward Mannock pointed out. First of all he was close on thirty years of age and secondly he had no sight in his left eye, two obstacles that under normal circumstances would rule him out. However, due to the fact that there was a great shortage of flying personnel he took his friend's advice and applied.

Meanwhile he read everything he could lay his hands on about flying. His Commanding Officer eventually backed his application for transfer and he was called for medical examination, one of the first requirements to become a pilot. The eye test was certain to fail him if things went ahead in the ordinary way but Mannock was lucky. He had to wait for some time in the doctor's office for the medical man to arrive and examine him. While there he memorised the eye chart and passed. It was to be the beginning of a new career.

He performed exceptionally well on the theoretical aspects of flying at the No. 1 School of Military Aeronautics and was posted to Hendon Flying School for training. His performance there at the outset was not very spectacular. Anxious to get the feel of an aeroplane completely by himself he got up very early one morning and unauthorised, took up a machine for a flight. He landed safely but later ended up before the Commanding Officer who had him grounded for some weeks.

Eventually Mannock was allowed to return to flight training and obtained his pilot's proficiency certificate on 28th November 1916. He eagerly looked forward to taking on his mortal enemy in aerial combat. He was commissioned Flying Officer on probation and spent time at Hythe Gunnery School and was later sent to No. 10 Reserve Squadron at Joyce Green on the Thames Estuary. There he met a highly decorated air ace Lieut. James McCudden who had come to instruct them. He became his great friend and greatly admired him. He was able to discuss with the air ace the pros and cons of the various fighter planes and also the war in which he was so anxious to get involved against the Germans. McCudden's family came from Carlow and so the Irish connection bound them closer together.

McCudden's job was to show the young pilots the tricks of the trade when in aerial combat and how, when and when not to spin the aeroplane. Mannock was very impressed and learned much about flying his DH2 machine. He had some hair-raising experiences. On one occasion he decided to put his plane into a spin at a much lower height than that advised by McCudden. His antics frightened all concerned and although he succeeded in what he had set out to prove, his behaviour was not appreciated.

A fellow pupil of Edward "Mick" Mannock at Joyce Green was Meredith Thomas who, many years later became Air Officer Commanding the Royal Air Force in India in the Second World War. "My first impression of Micky was that he was very reserved, inclined to strong temper but very patient and somewhat difficult to arouse. On short acquaintance he became a very good conversationalist and was fond of discussions or arguments. He was prepared to be generous to everyone in thought and deed but had strong likes and dislikes."

One of Mannock's great dislike was of course the German war machine. He was posted to Clarmaris near St. Omer in Northern France towards the end of March 1917 and on 6th April joined No. 40 Scout Squadron near Bruay.

A week later he flew his first mission escorting reconnaissance planes and came under ground fire. In the diary which he kept he made the following entry for 13th April 1917.

"I went over the line for the first time. Escorting FEs. Formation of six machines together. My feelings were funny. A group burst near me - about 100 feet. I did some stunts quite involuntarily. Lost my leader and deputy-leader but led the patrol down south. Returned safely after a very exciting time. ..."

It was a quick initiation into the hazards of being a fighter pilot. In his new career he saw the disadvantage he was operating under with limited eyesight. Because of insufficient experience he was often slow in reacting to attack and going on the offensive. Such things were not to the satisfaction of many in his squadron who, although never saying anything to him directly, were critical of his performances and felt he was afraid. He knew what they were thinking and sensed a certain amount of hostility. Sooner or later he would prove them wrong.

He was eager to shoot down his first Hun, at least it would prove to his colleagues that he was on his way. Although they did not know it, his eye problem had him at a disadvantage and he began taking every precaution to minimise this sight deficiency. He began to sight the guns himself, a job usually done by the armourer, fixing the sights finer than usual to help him when he would come in contact with the enemy in the air. This type of behaviour began to isolate him more from his flying colleagues who looked upon Edward as a loner. His own background as a boy growing up under harsh circumstances was in strong contrast to that of most of the other flying officers. Once while in the Mess in the course of some discussion he rasped "It's all very well for you fellows; you were born with a silver spoon. I had an iron shovel."

He nearly came to grief on 9th May 1917 when he got separated from two aircraft on patrol with him. Seeing his isolation, three German aircraft came to get him and when he went to fire his Lewis gun it jammed. Through quick thinking he avoided certain death by going into a vertical dive which he managed to come out of at 300 feet. He landed safely but visibly shaken. He still hadn't scored his first "kill" and was fearful of never doing so.

On 25th May Edward Mannock scored what was probably his first success when, following an attack on a German plane, it dived as if it was about to crash. Six days later he had a similar success but because he didn't see either of the planes crash he didn't claim them as victories. He was now beginning to be taken seriously by the squadron and was made a temporary flight-leader. He shot down another aircraft on the 7th June and entered it in his diary.

"I brought my first dead Hun down this morning - over Lille-North. Have been up to 21,000 in the morning (3.30 am) looking for early birds. Got rounds into a fat two-seater the other morning over Lens-Lietard. Sure I smashed him up, he went straight down without turning."

A week later he got two more. Mannock was now on his way and was gaining confidence in his ability. He almost crashed once when a piece of steel came off from his engine cowling and hit him in the right eye. He managed to make a rough landing but then collapsed. It was very serious for him to get such an injury but he recovered quickly following three weeks leave.

His homecoming was not what he had hoped it to be. His mother had been drinking hard for some time and it was no longer the household of his youth. She was aggressive towards him and he spent much of his leave with the Eyleses who really looked on him with admiration as if he was their own son.

He continued his success in aerial combat and on 19th July he was awarded the Military Cross. What he really thought of the actual award is difficult to assess and it only merited a one line entry in his diary.

Major Edward "Mick" Mannock, VC; DSO (2 Bars); MC.
Royal Air Force Museum, Hendon.

Later he was promoted 'A' Flight Leader. It wasn't the most popular award in the eyes of those colleagues who never could get close enough to understand him because he could be difficult. They also felt that many pilots had done more than Mannock but had not got any recognition for it. He had a chip on his shoulder towards those who had the advantage of a much better education and once in a discussion with a friend said "It isn't the school or the university, nor who your father is that matters, it's what you've got in your head and your guts." His outbursts on social snobbery were often an embarrassment to those who worked with him.

Outspoken at the way the war was being run, he was one of a group of pilots who expressed strong dissatisfaction with the fighter plane SE5A which replaced the very reliable Nieuport fighters and he didn't hesitate to point this out to the Air Chief Gen. Trenchard when he came on inspection to the squadron.

Mannock began 1918 by shooting down a German aircraft on New Year's Day. For some time he had been working extremely hard and under great pressure, taking the loss of any of his pilots very deeply. Following a spell of leave he was given a break from flying and assigned as a training instructor to 74 Squadron in Hertfordshire. In this job he was also very successful and became very popular with the young pilots under his care. There was a slogan directed at those young pilots which he had painted on the hanger walls governing air fights against single seater scouts which read *ALWAYS ABOVE; SELDOM ON THE SAME LEVEL; NEVER UNDERNEATH*. Many a young pilot later owed his life to Mannock's expert advice as a training officer. For example he said "remember that even good flying never beat the enemy. You must learn to shoot straight. It's one failing of some of our finest pilots. By the way, I advise you to sight your own guns. It's no use just leaving it to the armourer - he hasn't got to do the fighting."

He was promoted to the rank of Captain. When his instruction task was completed he was appointed leader of 'A' Flight Squadron which then moved to France. Mannock continued with his success as a fighter pilot when he resumed aerial combat.

His character was an extremely complex one. He could be generous, considerate and most unassuming yet there was a vicious streak to that character which surfaced when in conflict with the enemy. Against them he was ruthless. This was evidenced once when he engaged a German plane and shot it down. This wasn't enough for Mannock, he swooped low over the wreckage and seeing the two Germans still alive he sprayed them five times with bullets. Asked about this back at base he angrily retorted "The swines are better off dead. No prisoner for me!."

Yet beneath that ruthless fighting exterior there was another side. For example when a letter was dropped over the British lines from a German seeking information about his friend Fritz Frech - "4th September 1917 he fell between Vimy and Liever. His respectable and unlucky parents beg you to give any news of his fate. Is he dead? At what place found he his last rest?"

The letter which gave the address of the ill-fated German's parents was passed on to Mannock. It was he who had shot down the plane sending the pilot to a certain death. Mannock wrote the best letter he could to the bereaved parents and probably for the first time saw the other side of the war picture.

He scored four successes on 21st May and three days later he was awarded the Distinguished Service Cross. His tally of aircraft shot down at this time was forty. It was an extraordinary achievement for a one-eyed pilot. The secret of his success was his sheer determination to get the German enemy with whom he appeared to be obsessed. His commanding officer Capt. K.L. Caldwell had this to say of him:-

"Mannock was not a stunt pilot. I never saw him looping or wasting energy of engine power in this manner nor was he better than the average pilot. He really hated the Germans. There was absolutely no chivalry in him and the only good Hun was a dead one."

'Mick' Mannock was a hard task master and always drove himself to the limit. On any fight against the enemy in which he was leader he would never take chances and was always ready to try and get a colleague out of trouble in the air if he saw he was in danger. A pilot named Ira Jones under his command endorsed this when he said "Mick saved *my* life two days ago. My Vickers gun had a stoppage and at that moment an Albatross sat on my tail. I thought my number was up when suddenly Mick appeared from nowhere and in about five seconds that Albatross was finished."

By this time the enormous pressure under which he was operating began to affect him and he would often get very depressed. He hadn't many close friends with whom he could discuss things. There was one however, Sister Flanagan, a V.A.D. nurse on whom he was able to unload his worries and receive a sympathetic ear and sound advice.

He came home on leave and when he visited his old friends Mr. and Mrs. Eyles, they could not get over the change in him. He was no longer the bright and breezy Edward but was inward looking and morose, characteristics that would not be at all suitable for a fighter pilot. Mannock actually had a dread of being shot down. He disclosed this on one occasion to a few friends in 74 Squadron and while talking about it he actually broke down and cried. He was under enormous strain from combat. While on leave at the end of May he received a letter telling him that he had been promoted to the rank of Major.

The break did him good and when he returned to action he was given command of 85 Squadron which had previously been under the command of a famous Canadian Air Ace, Billy Bishop. He had shot down 72 planes and so his was a hard act to follow. As well as being promoted to the rank of Major, Mannock had also two bars added to his D.S.O. He made a great success of his command and inspired great confidence in those under him. His theme was "To fight was not enough, you must kill." With his pilots in the air he was extraordinarily generous and would often give them the chance to shoot down an aircraft even though he could have done it himself quicker.

Despite all his enthusiasm he was gradually becoming more morose. The loss of his great friend and hero McCudden who crashed while taking off on 10th July was a severe blow to him. Nevertheless he continued his endless pursuit of the enemy at every opportunity. By the end of July he had equalled the target of 72 planes shot down set by Billy Bishop. One more and he would be ahead.

In a flight over enemy territory on 26th July Mannock and Lt. Inglis set out with their two aircraft, the former hoping to corner a German plane that would enable Inglis have his first shoot down. Soon the opportunity presented itself when Mannock attacked a two seater enemy aircraft. He shot the observer and pulled away to let Inglis get in for the kill. However the enemy plane nose dived to the ground. Seeing it, Mannock broke the golden rule that he had always instilled in his young pilots which was never to follow the enemy down. In this instance he dropped down to around 200 feet to examine the wreck. Inglis followed. They then levelled off and headed for home. It was to prove a costly mistake for Mannock because his plane was hit by ground fire and plummeted to the ground. Tragically, Edward Mannock was killed. Although Inglis's plane was also hit he managed to make it back inside his own line.

A few days after Mannock's death, Lt. Callaghan of 85 Squadron wrote of him "Mannock is dead, the greatest pilot of the war. And his death is worthy of him." His grave was never found. Mannock acquired the name "Mick" in 40 Squadron. He was a complex character, of a rebellious nature, never liking the British class system and was a fervent socialist. While growing up he experienced poverty and worked extremely hard to better himself. Yet he was also an ardent patriot for Britain with an all-consuming hatred of the Germans. He appeared fearless but was in no way reckless. The greatest credit must be given him for overcoming the disadvantage of losing the sight of his left eye and becoming an enormously successful fighter pilot.

Major Edward Mannock displayed an old world charm towards girls. He dated them but nothing very serious ever came of it. Sister Flanagan attracted his interest but again nothing developed from it. As well as his obsession with Germany he was often scathing of the British High Command in their running of the war. The closing months of his life showed the considerable strain that he was under and one wonders why this wasn't observed more acutely by those in command. One afternoon in a conversation with his friend Lieut. Jones and two nurses, one of whom was Nurse Flanagan, the remark 'after the war' was passed. To this Mannock said "There wont be any after the war with me."

Sadly Major Edward "Mick" Mannock, DSO(2 Bars); MC; VC . lies in an unmarked grave. He was a gallant service man who distinguished himself in the field of aviation as a fighter pilot, becoming Britain's top flying ace of that First World War. Outside of military aviation circles he is little known. A wing of Canterbury Hospital was named after him and a block of dwellings built on the site of his old home carries his name.

On Armistice Day 11th November 1918 when toasts were being drunk in the Officers' Mess at No. 74 Squadron to all those members who had lost their lives, a special reference was made to "Mick" Mannock, 'the king of air fighters.'

The citation of the posthumous award of the Victoria Cross to Mannock in the Summer of 1919 read "For bravery of the highest order. ... This highly distinguished officer, during the whole of his career in the Royal Air Force, was an outstanding example of fearless courage, remarkable skill, devotion to duty and self sacrifice." Ironically the Victoria Cross was accepted on his behalf at Buckingham Palace by his father, the man who many years before had abandoned him and his family.

FROM SLUM TO SKY

Few if any achieved greater distinction in aviation following such a disastrous upbringing than American Jacqueline Cochran. She never knew her parents, having been handed over for adoption as a baby. Unfortunately her adoptive parents gave her a hopeless start in life. She grew up in a shack that had neither power or running water. She had little or no clothes and what there was of them were extremely shabby. She never owned a pair of shoes until she was eight years old and was more or less allowed to run wild. She received no guidance from her adoptive parents and received little education at a school going age. She had to go to work very early in life in a sawmill earning six cents an hour for a 12 hour shift. Her parents were engaged in this type of work in sawmill towns. There is some doubt as to the exact date of her birth but it was between 1906 and 1910.

Jacqueline Cochran was bright and intelligent and had no intention of staying at the bottom of the pile for the rest of her life. She left home in her early teens to better herself, getting work in hairdressing and beauty salons. She had an aptitude for this type of work and quickly gained experience. In a few years she was earning a comfortable living, the awful years of her childhood now behind her.

On the advice of a customer in a beauty establishment, she trained as a nurse for three years and then took a job with a doctor in a sawmill town. However, in her job she was again coming into contact daily with desolation, poverty and illness and was powerless to alleviate such matters without money. She left her nursing post and returned to the cosmetic and hairdressing business.

Jacqueline Cochran knew what she wanted and set out for New York with her talents. Ever conscious of her appearance, she presented a pretty picture to any prospective employer with her excellent grooming and long blonde hair. She sought a position with the famous Charles of the Ritz. She would be acceptable if she cut her long blonde hair. She herself would decide at anytime if she should cut her hair but would not be told so. She did not get the job but got one with Saks, a very fashionable salon on Fifth Avenue. From then on Jacqueline Cochran never looked back.

Very soon she was doing assignments for her employers in both New York and Miami and was also moving in the right circles. It was at a Miami party in 1932 that she first met Floyd Odlum who was later to become her husband. He was a millionaire many times over and also many years her senior. However, he was a solid, no nonsense, down to earth Methodist minister's son and from the beginning he and Jacqueline hit it off. He thought highly of what she had achieved to date and gave her sound advice. He suggested that she should take flying lessons in view of her constant travel. It was the best piece of advice that Jacqueline ever got and which would one day put her into the pages of aviation history.

She began taking lessons in 1932 and quickly picked up the basic elements of flying. She was in fact a natural flyer. When the time came for her to sit an examination for her pilot's licence she was given permission to answer the written part orally as her educational background had been poor. Nevertheless, she passed and soon hired a plane and flew to Canada. Further flying lessons were given her by a naval friend Ted Marshall and then she bought her first plane, an old machine for $1200.

Although very keen on aviation and now with a commercial pilot's licence, she also continued her work in the cosmetic and beauty business. In 1933 she opened her own beauty salon in Chicago and also a laboratory for allied products in New Jersey. To be able to fly was a great advantage to Jacqueline as she was able to reduce the enormous time spent on travelling between business locations. She had become very successful financially and replaced her old aeroplane with a new Waco. In a short time Jacqueline Cochran had come a long way from the slums of her childhood.

She was becoming firmly hooked on aviation and having gained considerable experience decided to take part in the England to Australia Air Race of 1934. Her friend Ted Marshall would be her co-pilot. Unfortunately, before the event he was killed in an air crash. Jacqueline flew to the funeral and brought her four year old niece along for the ride. She crashed on the way but neither of them was injured. She was lacking the expertise of instrumentation flying and took lessons to overcome this defect.

Competition flying had a great fascination for Jacqueline and firmly resolved to take part in the England to Australia race, she bought a Gamma aeroplane. At this early stage in her career Jacqueline was experiencing medical problems and shortly before the race she had to go into hospital for a stomach operation. She discharged herself without permission eight days after the operation to take part in the race

She had the Gamma modified to carry extra fuel tanks. Again she ran into problems. Twice the plane failed during the race, once when over the Arizona desert and again over New Mexico when she was forced to land. Flying the Gamma in the air race was also a type of business venture for Jacqueline. If she did well in the race, the plane's reliability would be highly publicised and of course would be very beneficial to Granville brothers, the manufacturers. She in turn would receive royalties on every Gamma sold. On the whole she found the Gamma unsatisfactory. She and her co-pilot again experienced engine trouble in Bucharest and were forced to retire from the race.

Following a four year courtship, Jacqueline Cochran married millionaire Floyd Odlum in 1936. Since she never knew who her real parents were, now that she was married she felt her husband had the right to know so she hired a private detective to find out. This he did and the information was given to her in a sealed envelope. She handed it unopened to her husband. He didn't open it either and it lay in a vault until after both had passed away when it was burned unopened.

Her husband was totally devoted to her and encouraged her all the way in her ventures. She had now become extremely well off because of her great business successes. She had bought 25 acres of land as an investment in South California before she married while Floyd bought 1000 acres. It was there the Odlums had their ranch, the Coachella, where friends and business people alike were invited. They entertained the highest society in the land on a lavish scale, from Presidents to five star generals.

Jacqueline and Floyd each had their own private secretary and personal maid. The ranch had an Olympic size swimming pool, a private golf course and its own telephone switchboard and operator. Many, however, found Jacqueline overbearing and bossy She was used to getting her own way and accordingly they gradually eased themselves away from her friendship.

As she became more and more involved with aviation, she constantly sought records to break . Although a fearless aviator, Jacqueline was a very responsible flyer. In 1938 she won the much prized Bendix Trophy Race, the first woman to do so. She flew a prototype Sikorsky pursuit aircraft and because of the extended tank capacity one of its wings was higher than the other.

It was to the Odlums Paradise ranch that Jacqueline's friend Amelia Earhart and her husband came to stay in 1937 before she set out on her tragic round-the-world flight. Jacqueline, who was credited with having some psychic powers sensed some tragedy ahead but failed to dissuade Amelia from undertaking the flight.

At the start of the Second World War, Jacqueline Cochran was among other civilian pilots grounded. She was disappointed at this as she felt she had a part to play in aviation and much to contribute to the war effort. Nothing further was achieved in her flying career until 1941 when she sought to ferry bombers from America to the UK where they were badly needed. The Minister of Aircraft Production in England was Lord Beaverbrook who supported her scheme. She also had the backing of American General Arnold. The American Air Force wouldn't condone the idea of any civilian pilot, even Jacqueline Cochran, of engaging in this type of war work. So they insisted that one of their own pilots would be on board to take off and to land and that is what was done.

She was very confident that an American air-woman had a role to play in the war and when she got to England she met Lord Beaverbrook and also Pauline Gower who was leader of the women ferry pilots in the Air Transport Auxiliary (ATA). When she returned to America she was invited to the White House by Mrs. Eleanor Roosevelt to discuss the matter and she also met Generals Olds and Arnold.

At this time, the American Ferry Command, later to become the Air Transport Command, was being formed. However, there were no plans to incorporate women into that command. Because of her experience and ability Jacqueline was invited to organise a contingent of women pilots and bring them to England to join the British Air Transport Auxiliary. When she came to England she was given the honorary British rank of Flight Captain. The first five American civilian women pilots arrived in the UK in April 1942 and were met on arrival by Jacqueline.

Despite her standing as an aviator, there was a certain amount of antipathy towards her because of the rank she was given and also her great wealth and influence opened doors to her which were firmly closed to her UK colleagues. For example, the American women pilots were based at White Waltham and Jacqueline was extremely kind towards them. She however didn't live on camp but stayed at the London Savoy Hotel.

In the Autumn of 1942 Jacqueline Cochran returned to America because women pilots were then being mobilised into the Women's Auxiliary Ferrying Squadron. In charge of this operation was another extremely capable woman aviator named Nancy Love and it was part of the Army Air Force's Transport Command. Jacqueline headed a training scheme for women pilots who had less than 500 hours flying time to their credit. The result was that the entire operation of both Nancy Love and Jacqueline Cochran became the Women Air Force Service Pilots (WASPS) in late 1943. The latter became Director of

Women's Pilots and Nancy Lowe was Executive Director on the staff of Air Transport Command (Ferrying Division.) Although involved with the military, the WASPS was in reality a civilian service and made a major contribution to the war effort. The group was disbanded in December 1944.

Meanwhile Jacqueline Cochran was forced to go into hospital again for two major operations. Quickly she fought back and her great undaunting spirit saw her through. Her fame as an aviator had put her on several aviation boards and she knew everyone that was worth knowing. She was one of the first Americans on VE Day to get down inside Hitler's bunker and had a trophy to prove it, a gold door knob off his bathroom door which she obtained by trading a packet of Lucky Strike cigarettes with a Russian soldier for it. She was the first American woman to land in Japan after the Second World War. She was in Tokyo on VJ Day and through her contacts with some generals on General McArthur's staff she managed to get herself aboard the battleship *Missouri* to watch the surrender ceremonies.

Widely known and respected, Jacqueline Cochran later flew to Shanghai where she was welcomed by Madame Chiang Kai-shek and presented with Chinese Air Wings in recognition of her work during the war. Although her record-breaking flights had been interrupted, Jacqueline fully intended to break some more records once things had settled down. Meanwhile she was proving a worthy women's aviation representative, flying to India, Palestine, Persia, Egypt and so on. As a Roman Catholic she flew to Rome where she had an audience with the Pope.

During the Nuremberg Nazi war trials Jacqueline visited Germany and gained access to those trials through Robert Jackson, Justice of the Supreme Court. In 1945, Jacqueline Cochran was awarded the Distinguished Service Medal. Despite her health problems she had engaged in an extensive flying schedule from the time the WASPS were disbanded at the end of 1944 until she went into training again for the Bendix Air Race. This type of competition, where records were there to be broken was closest to her heart.

She was out of luck in this race. Her aeroplane was an ex-government Mustang. Early on she had radio failure. Nearing the Grand Canyon she experienced very bad weather and decided to fly above it to a height of 30,000 feet. However, when she reached 27,000 feet her engine began to cut out and she was forced to descend again. To lighten her load she tried to jettison her external fuel tanks over the mountains but the dropping mechanism partly failed, leaving the rear of the tanks still attached and with the movement began to cause damage to the wings. She eventually landed safely at the end of the race and was only 6 minutes behind the winner.

Probably the biggest break in Jacqueline Cochran's aviation career came in 1947 when she met for the first time the great American Air Force pilot, Captain Charles 'Chuck' Yeager, shortly after he had become the first man ever to break the sound barrier in a Bell X-1 aircraft. Yeager had become an aviation legend even before then. When America entered the Second World War he became one of his country's great fighter pilots, shooting down 11 Messerschmitts on one flight alone. The meeting between Jacqueline and 'Chuck' Yeager was the beginning of a long and close friendship between the two families. Yeager found her full of interest in his flying career, especially in his recent breaking of the sound barrier.

He also found her volatile at times, bossy and domineering. However, she didn't intimidate him and they got on well from the time they met. Yeager and his wife Glennis were frequent guests at the Odlum ranch. He found Jacqueline to be besotted with flying and everything associated with it. After a short while he had got to know the disadvantageous background from which she came and he learned to admire how she overcame those terrible times and become not alone an extremely successful business woman but also a top aviator.

Early on she had said to Yeager "If I were a man, I would be a war ace like you. I'm a damned good pilot. All these generals would be pounding on my door instead of the other way round. Being a woman I need all the clout I can get." And she used that clout to great advantage especially where her aviation aspirations were concerned. Much of the clout came through her multi-millionaire husband Floyd who was the owner of numerous Companies including RKO, General Dynamics and the Atlas Corporation.

By the time Jacqueline Cochran got to know the Yeagers, sadly her husband was becoming crippled with arthritis. She was flying as much as possible, trying to break records wherever they presented themselves. Floyd was only too willing to back her projects, many of which cost a great deal of money. Her great ambition yet unfulfilled was to fly a jet aircraft and she set out to be trained. A Company of her husband's was then building the Sabre jet. Despite this however, it was two years before she was allowed to fly one under the aegis of being a specially appointed consultant for Air Canada.

Once Jacqueline Cochran got the confidence of being able to fly a jet plane, her next move was to try and break some speed records. She approached American General Vandenberg about this possibility and knew of course that Chuck Yeager would be the man to instruct her. A Canadian-built Sabre F-86 jet built by her husband's Company would, she felt, give more thrust for her record-breaking attempt. The record she had in mind was to become the first woman to fly faster than sound. As a civilian, she required special permission to use American Air Force facilities and equipment.

General Vandenberg looked into the matter and eventually permission was given. Yeager was chosen to be her instructor in the F-86. Writing about this many years later, Chuck Yeager recalled the arrangements he made with Jacqueline for her first training flight which was timed to take off at 6 am. She was requested to be at the base at 5 am to be briefed and to get her special flying suit on. Not at all used to taking orders, Jacqueline arrived at 5.45am. Yeager who had been on the field since 4.45 am brought her into his office and said

"Look, I want to tell you something. If you want to fly this program, you're gonna be here on time. You've got fifteen people out here working at four in the morning to pre-flight your aeroplane and get your gear ready, while you, a single pilot, can't get here on time. Look at all the man-hours you've already wasted for the Air Force, not to mention the guys who are busting their tails for you. If you want this program, you're gonna be here when you're scheduled to be here."

The dressing down worked and Jacqueline stuck rigidly to her training schedule. She had a master instructor in Yeager who, in one training flight, saved her life. While flying in another jet close by her, he saw fuel pouring out from her wing and on to the side of the fuselage. Immediately he spoke to Jacqueline by radio and gave her detailed instructions on what to do. He also contacted the Control Tower to have fire-fighting services on stand by. Under Yeager's instructions she made a perfect landing and was whisked from the cockpit in case of an explosion.

It was on the following day, 18th May 1953 that Jacqueline Cochran became the first woman to break the sound barrier. Confident of her ability she made her first supersonic dive from 47,000 feet. It went unrecorded so she repeated the performance. Thus twice in a very brief space of time she had gone through the sound barrier. To crown this, she and Chuck Yeager, the two fastest aviators in the world, for the benefit of Life Magazine and Paramount Films, climbed towards an altitude of 50,000 feet and in very close formation made a superb supersonic dive again. Jacqueline had every reason to be elated and

Jacqueline Cochran standing by a Northrop T-38 in 1962 Smithsonian Institution

Photo No. 72-6626

made the headlines of the world press.

Five years later she achieved further fame by setting new speed records in a Lockheed F-104 Starfighter. This was a Mach 2 machine (capable of twice the speed of sound), an aeroplane that many experienced pilots were scared of. Jacqueline proved to the world that she could master it.

In 1959 Jacqueline Cochran was elected president of the most prestigious international aviation organisation, the Federation Aeronautique Internationale. Its annual meeting was being held in Moscow. Some few years prior to this her husband Floyd had bought her a private passenger twin-engine Lockheed Lodestar aeroplane. She invited Chuck Yeager to accompany her on the flight as co-pilot and navigator. He was still an active service pilot but through her influence Jacqueline got him released. Actually it suited the American Air Force to get someone like Yeager into Russia and see as much as possible. He carried photographic equipment with him. Jacqueline, ever conscious of her appearance and of her position as a famous woman aviator brought along her maid, private secretary, hair dresser and an interpreter supplied by the State Department.

As president of the aviation meeting in Moscow she acquitted herself very well. However, she found too much red tape for her liking in the capital and didn't endear herself to many of the Russian Military who found her extremely demanding and arrogant at times. She was lucky to have Chuck Yeager with her to keep her as far as possible on the straight and narrow. When returning home she sought permission to leave Russia via Siberia, then over the Aleutian mountains into Alaska and on to her ranch in California. This was refused and she was directed to fly home via Romania. She landed in Sofia, Bulgaria and again had a war of words with the authorities as she demanded to be allowed fly on to Turkey from there. Again this was refused and she was obliged to fly on to Yugoslavia. From there she flew to Spain and on to Paris for the Air Show before returning home.

Jacqueline Cochran, although now no longer young, continued with her greatest love, aviation. Extremely wealthy, she could have easily now rested with her great achievements but she drove herself on. She became the first woman to fly a jet across the Atlantic, a Lockheed Jetstar from New Orleans to Hanover in Germany.

While she could be extremely demanding, Jacqueline was also very kind to her friends. When the Yeager family needed special medical treatment on a few occasions, it was she who arranged and paid for it. She thought very highly of Yeager, greatly admiring his flying skills and achievements. Because of this she didn't rest until she managed through her influence to have him awarded the Medal of Honour.

She herself in her long flying career had received every accolade worth having. Throughout that career she was obliged to undergo seven operations on her stomach and three on her eyes, yet they failed to stem her flow of energy. Eventually events began to catch up with her. A severe heart condition resulted in the fitting of a pacemaker and she was forced to give up competitive flying. Her husband Floyd, whose health had been impaired for very many years died in 1977 aged 80 . It was a blow that she never got over and found very difficult to accept. Soon after that she began to fail rapidly.

She was now a tragic figure. The great Jacqueline Cochran, friend of Presidents, Generals and Prime Ministers was coming to the end of the road. Because of her illness she sold the ranch and moved to a much smaller house. There she lived out her days, being forced to sleep setting up in a chair. In her last year or so she became extremely difficult and was unable to accept her great disabilities. Because of this, most of her friends apart from the Yeagers moved away.

Jacqueline Cochran died in 1980. She had risen from poverty and a deprived childhood to build up an extremely successful cosmetic business and through sheer dedication became the first woman pilot in the world to break the sound barrier. Her achievements were legion as were her friends and admirers. Yet sadly, when she died, only 14 people attended her funeral.

THE AIR CIRCUS MAN

Of all the aviators that have become part of history, Alan Cobham's name stands out for the major role he played in highlighting the enormous potential, the development of civil aviation could have on commercial life.

Cobham, whose father was in the clothing manufacturing business, was born on 6th May 1894 in south east London. He attended a private school for some time and had a one year stay at a local Council School. In 1904 he entered Wilson's Grammar School, an establishment that had been founded in the early 1600s. He was not exceptionally bright academically and this didn't bother him very much as his aspirations were to become a sheep farmer in Australia. He had relations there.

This did not materialise and at the age of fifteen he persuaded his father to allow him leave school and go into business. He was apprenticed for three years to a large firm of wholesale drapers named Hitchcock Williams earning ten pounds a year. He had been brought to see an airship flight at Crystal Palace when he was ten years old. It was a fascinating experience for the young Cobham. The interest in aviation had been sown and he began building large kites and flying them on Streathan Common with a pal. However, it was some time before a deeper interest in aeroplanes developed.

Alan Cobham had a great love of the countryside and although he was in business in the city he would take frequent trips into the country where he grew to love the farmer's life. He was beginning to climb the ladder in the wholesale business but didn't see it as being a lifetime career. He wanted to become a farmer.

He had an opportunity of training to be one through his father's cousin Donal Birchley who farmed at Brockbury. His parents were not happy with him wanting to leave business but eventually agreed to let him go. He said good-bye to city life and thoroughly enjoyed learning the trade of farmer. As time went by he began to see that unless he had capital or was able to raise some, he could never own his own farm. By 1913 his father's business was in trouble and so he had no hope of getting any financial help from him. He left farming and got himself another job in the City.

The outbreak of the First World War saw Cobham enlisting in the army. His experience of farm animals stood him in good stead when he joined the Army Veterinary Corps. He served in France and was promoted sergeant. While there his old interest in aviation was rekindled when he saw the participation of aircraft in many military sorties. He decided that he would join the Royal Flying Corps, if it were possible with the intention of making aviation a career. Up to then he had made enormous progress in the Veterinary Corps and his success with injured animals made him highly thought of. While home on leave he sought the assistance of a neighbour who was a civil servant in the War Office to get him an interview. Through him he was granted one and was accepted for training as a Cadet in the Royal Flying Corps at Hastings.

He spent six weeks there and acquitted himself very well. From there he was sent to Denham to undergo a preliminary course of ground instruction. Although he had left school ten years earlier at the age of fifteen, he got very high marks and in May 1918 was sent to Manston, in Kent for flight training.

Sergeant Alan Cobham was on his way and his aviation dream when he first flew kites was now becoming a reality. He underwent a very intensive flying course at Hastings and raised many a training instructor's blood pressure with some unorthodox flying escapades. Nevertheless, he was a natural flyer and quickly mastered his craft. Having completed his training as a Cadet on 17th August 1918, he was promoted to Flight Lieutenant and became a flying instructor. In this field he passed on to many students the skills he himself had quickly mastered.

The First World War ended with Armistice Day 1918 and as a consequence the demand for pilots in the defence forces was radically reduced. Alan Cobham, now twenty four, saw very little future for himself as a junior officer in military aviation but clearly saw the potential in civil aviation. He left the RAF and set out immediately to begin a new career. He was under no illusions regarding the difficulties he faced as there were thousands of young pilots like himself looking for openings.

After some abortive attempts in getting himself established, Cobham, through an advertisement in the *Aerial Register and Gazette*, eventually made contact with two brothers Fred and Jack Homes in Berkshire. Jack was a former RFC pilot who had been shot down during the war. The brothers, anxious to buy an aeroplane and charge to give pleasure flights were seeking an experienced pilot. Alan Cobham joined forces with the Homes brothers and formed a partnership mostly with borrowed money. From it Berkshire Aviation Co. emerged,

And so in May 1919 Cobham began a facet of his aviation career that in later years would make him a household name. Up and down the country crowds of people flocked to locations where flights would be available. Business was brisk and soon the Company had two aeroplanes. There was the odd mishap either on take off or landing but nothing of a serious nature. Cobham, apart from being a first rate pilot, was somewhat of a showman also and to help his business along he often sought the assistance of show business people who would be operating in the same town as himself. He would invite them along to the airfield from where he was flying thereby attracting further attention and publicity. It proved to be a successful idea in more ways than one. Having met the great comedian Will Fyfe in revue in Chesterfield, Cobham invited him along with some of his beautiful show girls to his flying field as a publicity stunt. Among them was a lovely girl named Gladys Lloyd with whom he fell in love and later married.

But it wasn't all a bed of roses. With the promise of financial support, Berkshire Aviation Co. established summer flying seasons at eight seaside resorts in 1920. The idea was a very good one but soon land owners, Councils etc. were proving difficult in supplying landing fields due to the risk of big insurance claims in the event of a serious accident on their property. Because of these difficulties the financial backers pulled out and the Company found themselves in trouble. Cobham had no spare cash, having put all he had into the partnership, including £250 borrowed from an uncle.

Reluctantly he pulled out of the Company and sold his share of the assets which consisted of an aeroplane and some engines. The money he got helped him to pay off some of his debts. He was now in a despondent mood and the bright future he had seen for himself in aviation was no longer there. He was out of a job and to survive he joined the 'rag trade' again There he began to do well but aviation was in his blood and soon he began to see that selling clothes was not for him and longed to be in the air once more.

Through a lady friend who had been an accountant at the Aircraft Manufacturing Co., later to be called 'Airco', Alan Cobham got an interview. The Company did aerial photography on a commercial basis and then had a vacancy for a pilot. He was accepted and within two days he was on his way to Manchester in a DH.9, accompanied by a photographer. It proved to be a wonderful opportunity.

As well as commercial photography, there was also a demand from private sources who wished to have their properties or estates photographed from the air. Prospects looked good and Cobham's brilliant flying which enabled his photographer to get some fine shots was of great benefit to the Company. However, at that time in aviation there was little in the way of permanency and eventually Airco went into liquidation. Although the outlook looked bleak, it proved to be a turning point in his aviation career and one on which he could look back in later years as the best thing to have happened..

Geoffrey de Havilland had been with the Aircraft Manufacturing Co. since 1914 as their chief designer. When it was known that the Company was to go into liquidation he left it and with the aid of four other brilliant men formed what was to become the famous De Havilland Aircraft Co. Then in its infancy, they approached Aerofilms to do all their aerial photography for them. Cobham in his earlier job with Airco had already done work for Aerofilms and they thought highly of him. The agreement was that the DeHavilland Co. was able to get the services of Cobham which in turn would give him the permanency that he sought and put him on the road to success.

Apart from photographic work Cobham ferried newspapers from one place to another, often beating competitors. The Company was also involved in air taxi work and he carried important personnel. 1921 was proving to be a successful year for him. At the opening of Parliament in Belfast by the King, Cobham, with a photographer, brought from London a specially printed edition of the Times for the occasion and returned that afternoon with photographs of the event for the following morning's paper. By now several papers used aeroplanes to obtain scoops over their rivals and Alan Cobham enjoyed every minute of this aspect of his work.

Another facet of his work was the delivery of de Havilland Aircraft abroad. In September 1921 he flew two DH.9C planes to Spain. The Company was rapidly building up business and at the same time he was building up his reputation as a thoroughly responsible and much sought-after pilot. With his future now secure he married Gladys Lloyd on 30th June 1922.

It is to Alan Cobham that a great deal of the credit must be given for the construction of the extremely popular de Havilland DH.60 light aeroplane. He had been asked by Geoffrey de Havilland what he thought the vital characteristics of a light aircraft should be. He replied that among other things it should be a two seater - the second seat for the pilot's girlfriend; it should have a range of about 350 miles so that the plane could fly from the UK to the Continent with an economical cruising speed of at least 80 mph and have space for luggage for two. From those salient points Geoffrey de Havilland and his team designed what was to become the famous DH.60 and which became the forerunner of the Tiger Moth.

Between 1921 and 1925 Alan Cobham spent much time ferrying a very rich American named Lucian Sharp around Europe and gained valuable knowledge of the likely civil air routes there. His employers de Havilland were fast becoming a very successful commercial flying Company. He was an ardent believer in the future of commercial aviation and always sought to publicise its advantages and future prospects wherever possible. He made newspaper headlines in September 1924 when he flew from London to Tangier in North Africa in thirteen and a half hours, displaying the enormous benefits to commercial life that aviation could bring. It was an uphill struggle for him at times but he persevered.

As a very experienced aviator with a fine record, Alan Cobham was much sought after for his advice in aviation matters. When Sir Sefton Brackner, Director of Civil Aviation was being sent to India from the UK to report on the feasibility of establishing mooring masts at Karachi and other Indian locations in preparation for airship flights, the Air Minister Lord Thompson had decided that Brackner travel there via the P & O line. Cobham saw in this a dichotomy so far as the British Government was concerned. For the Director of Civil Aviation to travel to India by liner on aviation matters wasn't the best way to promote air travel as the passenger transport of the future.

He pointed this out to Lord Thompson who explained that flying to India would cost twice as much as going by ship. De Havilland raised money to cover the cost of half the fare and the British Government paid the other half and so Sir Sefton Brackner was flown to India and back by Cobham. It was a marvellous scoop for Cobham and civil aviation in general. Even then he could see clearly that it was in aeroplanes that the future transport lay and not in extremely large and cumbersome airships. Flights such as that made to India and back opened up an aeroplane awareness that was otherwise slow in surfacing.

Although the de Havilland Hire Service Co. with which Cobham was deeply involved was operating very successfully, he could see that with technology advancing so rapidly there would soon come a time when newspapers etc. would be able to get their pictures over wire and by radio. Consequently the service being provided by the likes of the de Havilland Hire Service in getting photographs urgently from incidents abroad would soon be no more. In an effort to further promote the use of the aeroplane commercially he set about organising a flight to Capetown and back. This was an enormous undertaking and required great planning. It necessitated support groups along the route. Most of the flight would be through British Empire territory.

Sir Alan Cobham (right) during his Air Circus visit to Cork 1933
Courtesy Cork Examiner

De Havilland allowed him to have on loan the DH.50 in which he had flown to India. A more powerful engine was required and he was supplied with a 385-HP. Jaguar by Sir John Siddely. Cobham had several backers for this prestigious flight. Sir Charles Wakefield supplied the oil while B.P. supplied the petrol. The flight began on 15th November 1925, making numerous stops on the way. On board with Cobham were his great friend and outstanding mechanic Elliott and there was a camera man from Gaumont the film people who, on the instigation of Cobham, would film the entire undertaking. The flight had been very well publicised ahead of time and they got a rousing welcome everywhere they landed. At places such as Cairo, Johannesburg, Kimberly and Capetown people turned out in their thousands to welcome them. The publicity not alone for Cobham but for civil aviation was enormous.

The flight to South Africa and back was most successful and highlighted once again that the future of passenger transport would eventually lie with the aeroplane. As well as endorsing the merits of civil aviation, the film of the flight was highly successful and grossed big money. Alan Cobham's share was around £8000, a very large sum at that time. He also wrote a book on the flight called *My Flight To Capetown And Back*. On his return he was invited to Buckingham Palace by George V who had followed the course of the flight with great interest in the newspapers.

Cobham wasn't at all happy that the British Government was giving sufficient commitment to the aeroplane as being the commercial passenger carrier of the future. He was seeking ways in making them sit up and take more notice. Lobbying MPs and even Government Ministers did not appear to be enough. Perhaps if he could do something spectacular in the aviation field that would really grab the headlines, it might help.

He came up with a novel idea. He would fly to the House of Parliament by the most circuitous route and in a blaze of publicity. His idea was that, subject to de Havilland's approval, he would have the DH.50 in which he had flown so successfully to the Cape and back converted into a seaplane and then fly from Rochester to the Houses of Parliament via Australia landing on the Thames opposite the House. He would then deliver a petition to Parliament to take the aeroplane and its future in civil aviation more seriously. It was a brilliant idea but one that also required great ingenuity.

Cobham was not just a brilliant aviator, he was also an astute business man and a showman to match. His successful flight to the

Cape had enhanced his reputation. After his return from the Cape on 13th March 1926 he put his idea to Geoffrey de Havilland who fully endorsed it immediately. De Havilland supplied him not alone with the DH.50 aeroplane but also with secretarial assistance to handle the enormous amount of correspondence arising from the promotion and planning of the flight and in getting permission from the many authorities along the route to land on rivers, lakes etc.

There was then the question of the enormous amount of spares which were needed to be strategically placed along the route. A lot of spares remained following the Cape flight and it was arranged that they be despatched to places such as Singapore, Melbourne and so on. One of the great assets Cobham had on his many successful flights to date was the great assistance of his mechanic Elliott. An outstanding man at his job, his competence in keeping the aircraft fully serviceable at all times had proved to be an invaluable factor in Cobham's success. Both men held each other in very high regard.

On 30th June, about ten weeks following their return from the Cape, the two men set out again in their DH.50 now modified as a seaplane from Rochester via Australia to the Thames. One feature that stood out in the preparation of this flight was the strange feeling of depression that Cobham had. It had never happened to him before and it was a form of foreboding that all was not going to be well on the flight. And how right he was.

After some days flying along over the Euphrates they met several sandstorms on their way to make a landing at Bushire in the Persian Gulf. The weather got progressively worse and as they flew low to get their bearings over a swampy area to the north east of Basra, Cobham heard a loud explosion. A petrol pipe burst and his mechanic Elliott was crouched in the seat behind him in a pool of blood. They flew on to Basra where he was taken to hospital but sadly he died. It was discovered that a bullet fired from the ground had pierced the aeroplane and then a despatch box that Cobham was delivering from the Foreign Office to the Governor General of Australia. The bullet then went through a fuel pipe and through Elliott's arm and lungs. It was a tragic incident and Cobham was devastated.

Following an investigation the culprit who had fired the fatal shot was discovered among one of the Bedouin desert tribes. The incident was believed to have been of a political nature, the man who fired the fatal shot thinking it was one of the R.A.F aircraft being used at that time in the keeping of law and order. He was arrested, tried and sentenced to death which was later commuted to life imprisonment. About ten years later Cobham was given a different reason for the cause of the tragedy. It seems that following further investigation the authorities had decided that the man in question did not deliberately fire a shot with the intention of doing as much damage as possible to the aeroplane. He was a hunter who was about to take aim at a gazelle when out of the blue an aeroplane came flying very low and his immediate reaction was to fire at it in sheer desperation.

After the incident Cobham was in a dilemma, should he go on with the assistance of a new mechanic or abandon the remainder of the flight? He chose the former, having got the services of another top mechanic, Sergeant A.H. Ward loaned to him by the RAF. Other than that tragic incident in the early part of the flight, the remainder was an outstanding success and Australia was reached. There Cobham was again the hero and civil aviation got a further tremendous boost by his achievement.

The return journey went without a hitch. In his DH.50 Cobham had covered over twenty-seven thousand miles on the flight to Australia and back. It was on 1st October 1926 that he flew in over Westminster Bridge, landing on the Thames in front of St. Thomas's Hospital before a huge crowd of press and well-wishers. Quickly he was ferried across to the other side by motor launch and arrived on the terrace of the House of Commons where among those waiting to greet him was his wife Gladys. Also there were Sir Samuel Hoare, Secretary of State for Air, Sir Sefton Brackner, Director of Civil Aviation, Geoffrey de Havilland and many MPs.

Alan Cobham had once again made his point with regard to civil aviation, only this time in a much more spectacular manner. Press headlines on the following morning ensured the success of his flight and its objective. To honour his outstanding achievement he was again invited to Buckingham Palace where King George V conferred on him a Knighthood.

Sir Alan Cobham, a very happily married man and father of two children now felt that he had done enough publicity flying and that the time had come for him to go into business in civil aviation. He left de Havilland and formed his own Company with two other partners which was called Alan Cobham Aviation Ltd. in May 1927. The Company had been set up primarily as aviation consultants and as air route surveyors. Cobham had a wealth of experience in this regard but most of all he wanted to start up his own airline and had targeted Africa as being the ideal territory for this project. Imperial Airways had been flying as far as Cairo only and he had been told by the Director of Civil Aviation that there was no intention of the Company penetrating further into Africa.

Before he could make any inroads into the business of commercial aviation in Africa he felt it would be necessary to undertake a flight which would take him all over the African Continent to determine the most suitable routes, the establishment of aerodromes in conjunction with the various government authorities, the siting and location of fuel depots and the many other factors necessary before his airline could become a viable proposition.

It was an enormous undertaking. Early in his preparations Cobham discovered that the North Sea Aerial & General Transport Co. Ltd. had been thinking along the same lines. The Company had been formed by an old friend of his Robert Blackburn together with Captain Tony Gladstone. They met and agreed to pool their resources and in April 1928 Cobham-Blackburn Airlines Ltd. came into being. The survey flight over Africa was undertaken. It was not without many hair-raising incidents but it was a great success.

By now civil aviation was developing faster than heretofore and the potential of the aeroplane was being clearly seen. Imperial Airways, now a very successful airline decided that they would like to expand their aviation frontiers and began putting strong pressure on the UK Government to allow them to establish air routes through Africa. They succeeded in their effort. It was a David and Goliath situation and Cobham-Blackburn Airlines Ltd. could not compete and were bought out by Imperial. The Company was wound up in July 1930 for £24,750 and 25,000 Ordinary Shares of £1. It was a bitter disappointment to Sir Alan Cobham, the dream of owning his own airline was shattered but he had resilience and bounced back.

Probably one of the greatest contributions he made to civil aviation was in the first half of the 1930s when he inaugurated National Aviation Day. Its purpose was to make the ordinary people more air minded by bringing aviation to their doorstep in the form of an air circus. Such a circus would visit numerous places all over Britain and Ireland where there would be plenty opportunity for people to go up in an aeroplane for the first time in their lives at a very reasonable charge.

It proved to be an enormous success. Again great planning and organisation went into the project. Before the idea could be put into practice Cobham had to create an awareness among the many local authorities of the necessity to establish their own aerodromes. He called this aspect of the scheme his Municipal Aerodrome Campaign. Between May and October 1929 he visited some 110 towns and cities in a DH.61 aeroplane. These visits were well arranged in advance. In the first half of each day's visit he would take the Mayor, Councillors and other dignitaries for a flight over their city or town which indeed created a great impression. In the afternoons he brought groups of children for free flights. This he could afford to do due to the generosity of Lord Wakefield who had sponsored flights for ten thousand school children. Cobham's approach worked very well and soon aerodromes were laid by many municipal authorities, some of them becoming the forerunners of major airports in those areas today.

Cobham was now ready for his National Aviation Day, each town or city to be visited would have its air circus on its own special day. He put together a fleet of aircraft both small and large, among them being an Autogiro, Planette, Cowper Swift and a Handley Page W.10 which was billed as 'The Giant Airliner.' He had a team of first class pilots who carried fare-paying passengers and also pilots who were stunt flyers who performed daring feats of aerobatics. There was also a famous wing walker named Martin Hearn.

Between 1932 and 1935 Alan Cobham and his team gave countless thousands of people their first taste of flying. The air circus moved across Britain and Ireland, drawing enormous crowds wherever it went. The airship era had ended in England with the crash of the R101 and now the aeroplane was rapidly coming into its own. No one knew better than Cobham of its future potential in civil aviation and no one had played a greater part in highlighting this. National Aviation Day over a four year period was a great success and had laid a solid foundation on which civil aviation could be built. By 1935 however the air circus had run its course and Cobham sold the display to a noted aviator C.W.A. Scott, who with Tom Campbell Black had won the MacPherson Robertson Air Race to Melbourne in the previous year.

In 1936 Cobham opened a Company called Flight Refuelling Ltd. at Ford Aerodrome which he had actually formed in 1934. It was in response to an aviation problem that had been interesting him for a long time. A plane taking off on a long flight had to carry such a great weight of fuel that it restricted the weight of cargo and the number of passengers it could carry. Cobham realised that one way in getting over the problem was to refuel the aircraft in flight from another aeroplane called an air tanker. If this were possible there would be no need to take on board large amounts of fuel on take off thus allowing greater cargo and passenger carrying capacity.

There were several technicalities to be overcome especially the best type of fuel hose and how it could be attached to the other aircraft in the air. After much experimentation, Cobham and his team carried out extremely successful refuelling operations in the air.

It was now the era of the flying boat when bases were established at Foynes in Ireland and at Gander Newfoundland to enable Imperial Airways begin their Atlantic crossings. He had obtained from the airline a contract for inflight refuelling and the first refuelling was made on 5th August 1939 when the *Caribou*, piloted by Captain Kelly Rogers, was refuelled in flight off Shannon. A similar operation was undertaken off Newfoundland.

In all, sixteen crossings of the Atlantic were made using this technique. The outbreak of the Second World War put an end to them. The rapid advances in aircraft design after the war made this procedure obsolete for commercial aircraft but today military aircraft all over the world use in-flight refuelling techniques, albeit extremely sophisticated ones, to extend the range of their operations.

Cobham was the all-round aviator but his contribution to the development of civil aviation, especially in its early decades, was enormous. His death in 1973 saw the passing of a monumental figure in that field.

THE HINDENBURG

The German Count Ferdinand Von Zeppelin was first to see the great advantage of a type of balloon that could be steered, called a dirigible. He was born of a Prussian family in 1838 and went to America at the beginning of the Civil War where he joined the Union Army as an observer. Later, he saw action as a cavalry officer with General Hooker's Army at Frederickburg. After the war he joined an expedition to find the sources of the Mississippi and during this venture he took his first ride in a balloon. Quickly he saw the potential of such a balloon, especially if it could be guided.

However, it was some time before he was able to put his ideas to the test. When he returned home from America in 1866 he entered military service and reached the rank of Brigadier General on retirement in 1890. He was then aged 52. During the Siege of Paris he saw at first hand the great advantage of balloons for scouting purposes. On leaving the army he set to work on the design of a dirigible balloon . He had his design patented in 1894 and in 1900 the first dirigible airship the LZ1 was flown. It wasn't a great success and made only a few brief flights before being dismantled. However, it was the beginning of a long run of such airship designs known as Zeppelins. It wasn't until 1928 that one was officially given his name, the famous airship *Graf Zeppelin*. Many Zeppelins were used very effectively however by Germany during the First World War.

When matters stabilised after the war, the building of Zeppelins went ahead again. Their great originator, The Count, had died in 1917. The *Graf Zeppelin* was the most successful airship of all. History was made when the British *R34* made the first airship Atlantic crossing in July 1919 but the crash of the *R101* in 1930 spelled the end of the British involvement.

The Americans used the Zeppelin style airship for some time after that but following some crashes they turned their thoughts mainly to aeroplanes. Great strides were now being made in their development in many countries. America was very much to the fore, with for example the trial flighting in 1936 by the Douglas Aircraft Co. of their DC-3, a 21 seat passenger plane capable of a speed of 220 mph. The aircraft was capable of flying at an altitude of 20,000 feet with a heated cabin. There was one snag, the plane had a range of only 1,200 miles and so did not pose any threat to trans-Atlantic flights by airships such as the *Graf Zeppelin* and also the *Hindenburg* which was undergoing flight trials at that time. Seaplanes were expected to pose a threat but the Germans were not deterred, the extraordinary success of the *Graf Zeppelin* and now the *Hindenburg* coming on stream were powerful propaganda tools for a developing Nazi Germany.

America did not see airships such as the *Graf Zeppelin* and the *Hindenburg* as being an everlasting competitor to the aeroplane. Great advances had been made and while the DC-3 could not yet fly the Atlantic, sea planes were very much on the horizon for that route.

The *Hindenburg* was the pride of the German airship fleet. It had been designed by Dr. Hugo Eckner who had become Director of the Zeppelin Corporation in 1923. He was a very astute businessman whose guiding hand gained Germany enormous aviation prestige through the *Graf Zeppelin*. However, in the 1930s things began to change, the Nazis were taking control and Eckner disliked what he saw. He had to steer a careful course.

To build a new Zeppelin, the LZ129, later to be called the *Hindenburg*, a special Corporation was formed called the Zeppelin Airship Corporation in which the National Government of the Third Reich had a share. In fact one of the Corporation's directors was the Chief of Police, a man whom Dr. Eckner was obliged to accept in order to get sufficient capital to build the ship.

Work went ahead with the construction of the largest airship in the world and on 3rd March 1936 the first flight trial of 3 hours duration was successfully made. The airship had almost half as much capacity again as that of the *Graf Zeppelin*. With its 16 gas bags inflated, it had 7.2 million cubic feet of hydrogen which was capable of lifting a gross weight of 236 tons. It had a cruising speed of around 80 mph which gave it a fuel range of about 10,000 miles. This capability would enable the giant airship to stay in the air for five or six days.

During the first flight trial of the LZ129, the Mayor of Munich contacted Dr. Eckner as it was over the city and asked him what was the airship's name and he replied "Hindenburg". The name attracted press attention world wide but the Third Reich were not pleased with the publicity that Dr. Eckner and his team were getting and the Minister for Propaganda told him so in no uncertain terms. "We are the state now ... we are not only its laws, we are its achievements" and as Eckner was not a member of the Nazi party, he was not looked upon as being part of those achievements.

Following flight trials, the *Hindenburg* completed ten very successful flights between Germany and the United States and received a fantastic reception in America. By the end of 1936 the first regular passenger service began, the very wealthy being prominent on the passenger list.

The *Hindenburg* was going from strength to strength and its success was being availed of at all times by Hitler and the Third Reich for propaganda purposes - the Fatherland was leading the world in the air. The Olympic Games in Berlin in 1936 where world attention was focused was the ideal platform for such propaganda and the giant airship flew overhead during the opening ceremonies.

Dr. Eckner, who certainly was not a Nazi supporter, had his name deliberately kept out of the many press releases and publicity about the Hindenburg by the Third Reich. Despite all the pressure however, he continued to display his internationalism. He was extremely pleased that the United States was in negotiations with Germany for the building of two airships, one a 'Zeppelin'. In fact he had offered the plans of the *Hindenburg* and a sister ship LZ130 then under construction in Germany to the builders of the proposed American airships.

To lessen any influence he may be exerting, orders were issued by the Propaganda Ministry to the Reich League of Periodical Publishers :-

"We request our members to bring to the notice of their editorial and writing departments the following instruction and ensure its unconditional observance:

The name of Dr. Hugo Eckner will no longer be mentioned in newspapers and periodicals. No pictures or articles about him shall be printed. We learn that the reason for this is the very strange stand assumed by Dr. Eckner."

The first trans-Atlantic flight of 1937 by the *Hindenburg* to New York was scheduled for 8 pm on Monday 3rd May. There was a great buzz of excitement among the Zeppelin crew as the German flagship of the air was about to begin a season of 18 trips to the US. Surrounding all this was the German political situation, at the heart of which was the Third Reich which was slowly but surely tightening its grip on everything German, the Swastika becoming the flag. Hitler, Minister of Propaganda, Goebbels, Goering, the head of Air Ministry etc. were now in control. There were very many in Germany of course who didn't like and feared what was afoot, especially the Jews.

There were great fears among those in high places in the Third Reich that an attempt would be made to sabotage the *Hindenburg*. The security branch of the Gestapo which had been gathering intelligence bore this out. It would not be the first attempts at sabotage. In March 1935 a bomb had been discovered under a table in the dining room of the *Graf Zeppelin* and rendered harmless. In 1936 an unidentified passenger travelled on the *Hindenburg* on a Swedish passport and landed at

Frankfurt. The Gestapo had him under surveillance and managed to search his hotel while he was out and found plans of both the *Hindenburg* and the *Graf Zeppelin*. The man never came back to his hotel, having probably been tipped off. There were several other incidents which indicated threats of sabotage.

Now, on the eve of the departure of the *Hindenburg's* first Atlantic crossing for 1937, the Third Reich were extremely worried and security was very tight. The entire airship was thoroughly checked by state security. They had every reason to believe from intelligence gathered that an attempt would be made to sabotage the airship when it was over American territory. For many months the German Ambassador in Washington had been receiving phone threats to the *Hindenburg* and as the persecution of the Jews and also of many Roman Catholics had intensified, their relatives in America and elsewhere were suspect.

A month or so before the *Hindenburg* was to make its first 1937 flight, the German Ambassador in Washington received a letter from a Mrs. Rauch informing him that saboteurs intended to plant a bomb on the airship She urged the Ambassador to advise the Third Reich to cease the airship flights to safeguard human life.

So there was a major security alert and as the crew prepared the giant airship for lift off, the utmost precautions were taken to ensure that nothing untoward would happen to the airship. The passenger list had been scrutinised, and the backgrounds of those travelling investigated as far as possible. As an added precaution all passengers and their luggage were thoroughly searched and items such as cigarette lighters, matches, flashlight and photographic batteries were taken and placed in marked bags to be returned at the end of the voyage.

As an added precaution, the passengers for the *Hindenburg's* first 1937 flight were brought to the airfield by buses. As they filed on board, a brass band played some traditional music, ending up with Deutchland Uber Alles. At 8.15 a.m. on 3rd May 1937 the airship left Rhein-Main airport to the cheers of the ground staff and the many relatives of the crew who had come to see them off. A detachment of the Hitler Youth Movement were also in evidence to bid bon voyage to this enormous airship. In overall command was Captain Lehmann. There were three captains to cover the three 8 hour watch period and an overall crew of 61, many of them trainees who later would be deployed in the new airship being built, the LZ130 - Graf Zeppelin II. The *Hindenburg* was the biggest airship in the world.

The airship passed over Cologne where mail was dropped. Because of the political situation, the *Hindenburg* avoided going over French and

English territory. Its course was plotted along the Rhine, then over Holland and south west through the English Channel and out over the Atlantic. By Tuesday morning, when the passengers came for breakfast, the airship was a considerable distance over the sea. They were delighted with their luxurious accommodation. For many, especially the 'bad sailors' it suited them far better to travel by airship than on the luxury liners then afloat. Their flight was calm and smooth and they could sit in comfort and listen from time to time to the ship's news on the wireless. One of the main topics was the impending coronation of George VI and the divorce of Mrs. Simpson. A weather report indicated a small depression on their track but it was hoped to avoid it.

To entertain the passengers, conducted tours were provided. Each passenger was supplied with sneakers to avoid creating any sparks that could make a potential fire hazard with the hydrogen. An interesting feature of the *Hindenburg's* luxury travel was the very high quality of food and drink for those on board. There was a magnificent wine list. Before the airship left Frankfurt there was put on board some 5500 pounds of fresh meat and chicken, 440 pounds of butter and cheese, 800 fresh eggs 220 pounds of fish and there was caviar and a large supply of salad. Every passenger's needs were fully attended to in this type of luxury travel. The exception of course was that they were much more confined to their quarters than the freedom on the decks of a luxury liner.

The passengers did not concern themselves very much with the fact that the fuel hydrogen which kept the giant airship aloft was highly inflammable. Helium, the much safer gas and very much favoured by the Americans in its airship construction plans was enormously expensive and if used would put the cost of airship travel too high. It was the extreme care exercised by Zeppelin crews in the handling of hydrogen that had made them so very successful to date.

Wednesday 5th May saw the *Hindenburg* much closer to its destination although the weather had slowed its progress. It would not now arrive at 0600 on the following morning but twelve hours later. It was a disappointment to all on board and also to those anxiously awaiting their arrival. The American radio announcer Herbert Morrison's broadcast of the *Hindenburg's* arrival and his intended interviews with the passengers had been thrown off schedule.

As Thursday 6th May dawned the airship was nearing the end of its flight from Frankfurt to New York. Flying at a height of 700 ft., west of Yarmouth and about 350 miles (563 km) from Boston, the *Hindenburg's* passengers came to breakfast and had their packing and the anticipation of meeting their loved ones to think about. Eventually Boston came into view and they could see from a height of around 500 feet the people on the streets below. Later New York hove into sight. To the delight of all concerned Captain Pruss brought the airship over the city.

Margaret G. Mather, a passenger, described this part of the flight :
"We flew over the Bronx and Harlem, then along Fifth Avenue, past Central Park, then we turned west and flew over the *Rex* and other big ships, down to the Battery. There we swung around to the East River, flew over two or three bridges, then across Times Square and out to New Jersey. The clouds were black and ominous as we flew over Lakehurst. The landing crew was not there and the weather was becoming worse instead of better so we flew on to the coast and cruised up and down along the beach, sometimes out to sea."

All this time the *Hindenburg* was being furnished with up to date weather reports. As the weather deteriorated and a report received at 17.30 pm recommended that the landing be delayed, the captain confirmed that he would await an improvement in the weather before attempting to land. At 1812 the *Hindenburg* received a message "Conditions now suitable for landing. Ground crew is ready, thunderstorm over station, ceiling 2000 feet." It looked a rather contradictory message - on the one hand it was saying that it was safe to land and on the other there was a thunderstorm over the airfield at Lakehurst. The airship however set course for the airfield and at 1822 another message was received by the captain "recommend landing now commanding officer."

Around 1900 as the *Hindenburg* approached the field the crew were summoned to their stations and at 1919 Captain Pruss gave the order "All engines astern." Below, the huge ground crew awaited the dropping down of the many cables to make the airship fast. As the landing procedures went ahead, Herb Morrison was doing his live radio commentary for WLS, Chicago.

"Here it comes, ladies and gentlemen and what a sight it is, a thrilling one, a marvellous sight. ... The sun is striking the windows of the observation deck on the westward side and sparkling like glittering jewels on the background of black velvet."

It was then that it happened. Inside the giant airship a crew member named Helmet Lau saw a bright flash towards the front around the area where Gas Cell IV was located. Quickly a small fire started and in an instant there was an explosion followed shortly by a second. As

Morrison was going ahead with his commentary he broke in with an excited voice:

"Oh! oh! oh! Its burst into flames ... get out of the way please! It is burning, bursting into flames and is falling ... Oh! This is one of the worst ... Oh! It's a terrific sight.. Oh! ... and all the humanity..."

By now the *Hindenburg* was burning rapidly and the ground crew underneath the airship ran for their lives. There was confusion and screams everywhere. Then the airship hit the ground and in a short time it was a raging inferno as the 7 million cubic feet of hydrogen burned in a ball hundreds of feet into the air. At first sight it would appear that no one could survive from within the airship. Many jumped from windows of their cabins with their clothes on fire. A cabin boy named Werner Franz who was only fourteen years old on hearing an explosion ran for the gangway and fell due to the severe tilting of the airship. There was fire all about him and in a few seconds he too would be burned alive By a stroke of good fortune a water tank burst above his head drenching him and he managed to jump to safety through a hatch to the ground.

Of the 36 passengers who were on board 16 died either in the airship inferno or later in hospital. Seventeen crew members also died. There were many who received very serious burns but who lived to tell their extraordinary stories.

Among the passengers who survived were German writer Leonhard Adelt and his wife. He had earlier collaborated with the Captain of the *Hindenburg* - Captain Lehmann in writing a book "Zeppelin - The Story of Lighter- than- Air Craft." Adelt painted a graphic picture of the last moments of the *Hindenburg* and the lucky escape of himself and his wife.

"... With my wife I was leaning out of a window on the promenade deck. Suddenly there occurred a remarkable stillness. The motors were silent and it seemed as though the whole world was holding its breath. ... I could not account for this. Then I heard a light, dull detonation from above, no louder than the sound of a beer bottle being opened. I turned my gaze towards the bow and noticed a delicate rose glow, as though the sun was about to rise. I understood immediately that the airship was aflame. There was but one chance for safety - to jump out. The distance from the ground at that moment may have been 120 feet. ... but in the same instant the airship crashed to the ground with terrific force. Its impact threw us from the window to the stairs corridor. The tables and chairs of the

The Hindenburg in Flames   Zeppelin Museum, Friedrichshafen, Germany.

reading room crashed about and jammed us like a barricade. "Through the window" I shouted and dragged my wife with me to our observation window. ... I do not know and my wife does not know how we leaped from the airship, the distance from the ground may have been 12 or 15 feet. ... All at once I had a feeling that my wife was no longer by my side. I turned about and the flames and poisonous smoke vapours struck me squarely in the face. I saw my wife stretched out full length and motionless on the ground. I floated to her and pulled her upright. I gave her a push and saw her running again like a mechanical toy that had been wound up. ..."

After a desperate struggle Leonhard Adelt and his wife reached safety. Others were not so lucky.

Ironically on that fateful day 6th May 1937 Colliers Magazine carried an article by writer W. B. Courtney regarding a trip he had made on the *Hindenburg* the previous year:

"... German dirigibles since (the First World War) have flown nearly one million passengers without a fatality. And it is the firm conviction of this sceptical reporter, after close first-hand watching of their methods, that only a stroke of war or an unfathomable act of God will ever mar this German dirigible passenger safety record."

Dr. Eckner, the designer of the airship on hearing of the disaster stated to reporters that it was probably caused by sabotage and went on to say"

"I have repeatedly received threatening letters, specially warning me not to land the *Hindenburg* at Lakehurst. It is quite impossible that the airship's explosion was due to lightning as it was equipped with the most modern preventive apparatus."

In a later interview Dr. Eckner modified his remarks to only "a slight chance of sabotage" perhaps at the instigation of the Third Reich to whom the loss of the Hindenburg as a form of aviation propaganda was an enormous setback.

Shortly after the disaster, the American Department of Commerce under Secretary Daniel C. Roper set up a Commission of Inquiry into the cause of the explosion. Among the Germans who joined as 'observers' was Dr. Eckner. The Commission sat for 6 months. Numerous theories were put forward including that of an experimental rocket having been fired by an inventor in Connecticut and hitting the airship.

Aviation experts were all agreed that, discounting the rocket theory which they looked upon as not practical at that time, it was just not possible to ignite the hydrogen inside the airship. Sabotage on board of course was on the minds of many but perhaps the Americans did not want to put forward that idea in case they got involved in an international incident on their territory. Accordingly they "would not consider sabotage" and eventually their findings stated that among the most probable causes was "St. Elmo's Fire." This was an electric discharge sometimes seen on ships masts and on the aerials and wings of aircraft. A noted airship historian (Dr. Robinson) later interviewed a witness to the disaster who testified to seeing the St. Elmo's Fire on top of the airship before the fire occurred. It would appear however, that this was the only sighting of this affect. There had been a special confidential report on the suspected sabotage theory prepared by the German Technical Commission which was given to the enquiry but the Third Reich had no intention of admitting in it that the *Hindenburg* had actually been sabotaged. Goering, the head of the Luftwaffe declared that it was "an act of God."

The *Hindenburg* tragedy apart from looking at it in a national or international context was an enormous loss to all those involved with the Zeppelin Company. The wives and families of practically all the crews of the airships lived at Frankfurt-on-Main. Many young girls had come there with their husbands when the Zeppelin headquarters were moved from the shores of Lake Constance to Frankfurt. New homes were being built to house young families in the special "Zeppelin Village."

Ironically, scrap metal from the wreck of the British Airship *R101* which had crashed in France in October 1930 had been purchased by the German Government. The twisted framework had been shipped from France melted down and used in the *Hindenburg's* construction. German design engineers felt however that by mid 1936 they had introduced the safest engineering features into Zeppelins making them "as safe as trains". The *Hindenburg* described as a "grand hotel of the air" was 815 feet long, almost as long as the *Queen Mary*. The envelope of the airship had been treated with an aluminium powder to minimise the effect of heat rays from the sun.

There were several other theories as to the cause of the *Hindenburg* disaster . Dr. Eckner, the designer of the airship and who earlier had made comments regarding sabotage, was later to say that he believed the last turn the ship took before docking may have put excessive stress on the hull, snapping a bracing wire which cut into gas cells, thereby leaking hydrogen which caught fire.

The celebrated author Len Deighton in his book *Airshipwreck* tells of a conversation he had with Captain Hans Von Schiller in 1974 and who at the time of the *Hindenburg* disaster was Captain of the Graf Zeppelin on its way back from South America. Von Schiller commenting about the cause of the tragedy said:

"No one can say. Pruss and Lehmann were the finest airship commanders, men of vast experience. But maybe, I say maybe, when they were landing, a tail wind blew back along the ventilation. There were always a few leaks of hydrogen. Air blowing from behind would prevent it escaping. Now, when an airship lands it reverses its engines, just as a ship does. And, like a ship, it vibrates. Now I'm not saying this happened, I am saying maybe. But if the zeppelin vibrates so much, it could break a wire. These bracing wires are thick, and when they snap, heat is generated at the break. Now if the hot end of the wire recoiled into the build-up of leaking gas, this could have caused the fire."

The *Hindenburg* disaster spelled the end of airship travel. The *Graf Zeppelin* was the last to carry fare-paying passengers. On 18th June 1937 the *Graf Zeppelin* flew from her home base at Friedrichshafen to Frankfurt where the airship was deflated and used as an exhibit there. Helium, by far the safer gas, would have not been practical for giant airships like the *Hindenburg* due to its poorer lifting capability than hydrogen and its very high cost. A political decision in America had at any rate blocked the export of helium to Germany.

In his fine book *The Hindenburg* published in 1972, the author Michael M. Mooney, having undertaken exhaustive research both in Germany and in the United States, leaves us in little doubt whatever as to what he thought was the cause of the *Hindenburg* disaster - it was he says a deliberate act of sabotage by explosion. He found the US National Archives to be an extremely revealing source of information on the sabotage issue. He even discovered data describing the materials from which the bomb was made as analysed from the remains collected at the scene of the disaster by Detective George McCartney of the Bomb Squad of the New York Police. Referring to this data he wrote:-

" ... I could read the eye-witness account of the bomb's explosion inside the ship given in public testimony by Helmet Lau and then turn to private diaries, papers and correspondence and see how the bomb was made. Instead of being unable to discover sufficient details, I soon had too many from a combination of officially recorded evidence and unofficial, but now available materials. ..."

The planting of an incendiary device was always a fear in the minds of the German Authorities and the senior officers of the *Hindenburg*. One point in favour of the sabotage theory was the fact that the bomb was timed to go off some time after seven pm on Thursday evening. The Hindenburg would have landed under normal circumstances at 6 am and so would have few personnel on board when the bomb went off, thus avoiding the enormous casualties that did occur due to the postponed landing.

While the Hindenburg had been flying a sister airship the LZ130 was being built and when the *Graf Zeppelin* was taken out of service, the new ship was called *Graf Zeppelin* also. It never carried fare-paying passengers and flew only about thirty times in all. In fact the German Luftwaffe in 1940 under Goering ordered the destruction of both Graf Zeppelins. The airships had outlived their usefulness and went into history.

Despite so much documentary evidence that advances the sabotage theory on the destruction of the *Hindenburg*, it is extremely doubtful if this will ever be fully confirmed by either Germany or America .

ROMANCE IN THE AIR

One of the great aviators of the 1930s was Jim Mollison, a rather flamboyant individual who, apart from his great flying achievements, attracted considerable publicity from his marriage for some years to the celebrated aviator Amy Johnson.

He was born in 1905. As a young man, his mother felt that it was best for him to join the RAF. He had never passed an examination during his school years and was asked to leave Glasgow Academy at an early age. However, through the connections that his mother and grandfather had he got into the RAF on a Short Service Commission.

The Service didn't suit Mollison's temperament as there was too much military discipline. On one occasion during his training he overstayed his leave and was court-martialled. On his second solo flight he crashed in an area where he shouldn't have been at all and caused considerable damage to his aircraft, nevertheless, he was a fine flyer and after some time he steadied down and passed with distinction as a pilot. Many years later he was to admit that he managed to get hold of a copy of the examination questions. He was commissioned as the youngest officer in the RAF at 18 years and 3 days.

In due course Jim Mollison was sent to India and was on active service in Waziristan in February 1925. This qualified him for a General Service Medal. He did not enjoy the Air Force regime although the training and discipline he received stood him in good stead and he very quickly became an excellent pilot. When he returned to England he was sent on a gunnery course which he failed to pass. It just wasn't his cup of tea.

At the age of 22 he became a qualified flying instructor and in the space of 14 months had taught 15 young officers to fly. It was however apparent to his superior officers that Service life wasn't for him and on completion of his Short Service Commission he was transferred from the active list to the Reserve as Flying Officer Class C on 14th March 1928. He came out to civilian life with a gratuity of £350, a great sum then.

Mollison had a considerable attraction towards the opposite sex and with money in his pocket he decided to make up for the years he had spent in uniform and travel to the south of France to soak up the sun and meet the girls. He fell in love while in the Riviera with a girl named Paula and got himself engaged. Luckily they decided that it was better they wait until he had got himself a proper job before they married. Having lived it up for some time he and Paula parted and he moved on.

Mindful of the fact that his money was slowly diminishing he took a ship to Australia via Tahiti. In Tahiti he stayed in blissful happiness with another beautiful creature named Toi for some time before eventually setting off again and eventually landed in Sydney. By this time, most of his money had gone and he needed work very badly.

He took a job as a bathing-beach attendant on a very meagre salary. One day an aeroplane from the New South Wales Flying Club few low overhead. This immediately re-kindled the flying bug in him and he applied for a job as an instructor at the Club. There was no vacancy but he was advised to try the South Australian Flying Club. He was taken on there as an instructor. He didn't have a "B" licence which was a legal requirement to be an instructor in civil aviation although the Club didn't appear to realise at the time that it was necessary.

Jim Mollison quickly took to the air. However, after a week as a flying instructor a telegram was received by the Flying Club from the aviation authorities which said "Stop Mollison flying". He was determined to fight this obstruction and travelled to Melbourne in November 1928. There he met Colonel Brimstead, the Aviation Chief and put his case forcefully to him. Although from a legal standing, it was perfectly correct that Mollison could not be a civil aviation instructor without a "B" licence. Brimstead, a far-seeing man realised that commercial pilots were very scarce in Australia and had Mollison issued with his "B" licence without sitting any examination, much to the latter's relief. He returned to Adelaide for a year where he was able to certify thirty three new pilots.

He was anxious to move on to better things and he left his job to work for the famous aviator Charles Kingsford Smith who gave him a pilot's job in Australian National Airways with the rank of Captain. The two men became great friends. Tragically Kingsford -Smith died in an air crash in 1935. Writing about him, Mollison said "With 'Smithy' I flew many hours as co-pilot in his famous old 'Southern Cross'. I considered him the greatest pilot in the world and I have found no one his superior since he died."

Among the many people that Mollison met while working for Kingsford-Smith was Amy Johnson. He had been flying a special charter from Brisbane to Sydney and she was a passenger. She was already a big name, having flown solo from England to Australia. Mollison and herself chatted a lot about flying but did not meet up again for about two years,

Jim Mollison always had great confidence in his own ability as a pilot and had the secret desire to break records and gain prominence in the world of aviation as his employer Kingsford-Smith and others had done. Sponsoring was the main requirement to undertake record-attempting flights and Mollison as a £15 a week pilot saw little hope otherwise. He had discovered that many of the flying ventures to date had been backed by the oil baron Lord Wakefield. Mollison had often flown Cyril Westcott, Wakefield's manager in Australia on business trips and he put his ideas to him. He felt, he said, that he could fly equally as well and as quickly as any of the record-breaking aviators and he asked Westcott if there was any chance of Lord Wakefield helping him to attempt the Australian-England record.

Westcott succeeded in getting the backing for Mollison and a new plane was built for him for that purpose. When all was ready he flew it from Sydney to Darwin, his starting point, in May 1931. He refuelled there, and took off but unfortunately struck a telegraph pole, wrecked the machine but escaped unhurt. He was devastated and thought that it would be the end of any support he would ever get again. The incident however attracted enormous press coverage and Mollison received big publicity as did his benefactor Lord Wakefield. It gave Mollison renewed hope and he pleaded with Lord Wakefield to give him a second chance. He was successful.

Jim Mollison was now on his last chance. He already had resigned his job with Kingsford-Smith and had only a few pounds in his pocket. This time he made no mistake. The delay however meant that Scott, another great aviator, had gone on ahead of him crossing to England in a record-breaking flight. Mollison justified Lord Wakefield's confidence in him because he landed in England roughly a day and a half inside Scott's time.

When Mollison landed at Croydon, he received an enthusiastic welcome and was now a record-breaking aviator in his own right. As a result, he was offered a suite to rest in after his epic flight by the Grosvenor Hotel and a very large cheque from Lord Wakefield awaited him. Also a gentleman named Whitelaw whom Mollison had never met and who later became a good friend wrote offering him a thousand pounds for his great achievement. Mollison was in the money once more. Best of all he was offered £200 per week and expenses by the Daily Mail to fly an aeroplane on an instructional tour of Scotland which he readily accepted. In all, his flight from Australia to London had netted him about £7000.

Mollison was a born flyer. Among his criteria to be an "ace" pilot in the 1930s were:

(a) One should possess a disposition to be able to persuade other people to buy aeroplanes in which one could fly to fame.

(b) One needed to have complete disregard for one's own life and comfort and that of the crew if on board. At the same time one should plan and prepare as far as is humanly possible to see that the necessity for dying doesn't exist.

(c) One needed to have courage to disregard and throw away prospects of regular employment.

(d) One must have that 'something' to attract the public and media interest if one achieved something.

At a cocktail party in London in the Spring of 1932 Jim Mollison was introduced to an extremely attractive eighteen year old girl named Lady Diana Wellesley. She was a great grand-daughter of the Duke of Wellington. He was aged 26 and was something of a celebrity due to his solo flight from Australia to England in August of the previous year. The couple spent a considerable amount of time together. Soon the matter became public and to prevent any sinister tongues wagging they decided to become engaged.

There was strong opposition to this especially from Lady Diana's mother, Countess Cowley. Mollison at the time was preparing for his first flight to the Cape from Lympne Aerodrome near Folkstone. Great interest had arisen about this proposed flight and to add to the publicity placards blazoned "Jim Mollison engaged", "Mollison to wed society beauty" etc. were to be seen at the aerodrome. The engagement had not been officially announced and the Countess was livid.

Leaving all the speculation behind him, Mollison set off for the Cape. However, Lady Diana's mother contradicted the matter in the press by saying "Reports of Mr. Mollison's engagement to my daughter are untrue. My daughter is barely eighteen and far too young to think about marriage or an engagement. I would not consider the matter for a moment."

When Mollison returned from the Cape he met Lady Diana and they decided to tell the Press that the engagement was on a trial basis for a year, after that the couple would know their own minds better. It probably was an easy way out of all the publicity and no doubt Countess Cowley, Lady Diana's mother saw little future for her daughter with Jim Mollison who by then was giving more attention to Amy Johnson.

He had met up with Amy again in Cape Town when he had made that flight from Lympne in a D.H. Puss Moth in 4 days and 17 hours. She congratulated him and they went to some parties together. Two months later, with both of them back in London, he asked her to lunch one day and more or less on the spur of the moment he asked her to marry him. She laughed at him and then said "I'll take a chance." It was not the soundest of foundations on which to build a marriage that took place on 29th July 1932. They were poles apart in most things except aviation. Nevertheless, Amy Johnson the very practical business woman as well as famous aviator had fallen for the impulsive, restless and smooth-talking Mollison but who was also of course an outstanding aviator.

The whirlwind courtship and wedding attracted enormous publicity. After a honeymoon spent in Kelso Castle in the west coast of Scotland, Mollison's itching wings again took over. Before his marriage he had completed plans to attempt a solo East-West crossing of the Atlantic. Now he began to set those plans in motion. First he flew his single-engined D.H. Puss Moth from Stag Lane aerodrome to Dublin where from Portmarnock Strand he hoped to take off with a full tank of fuel. The weather was bad for the time of year and he had to wait in Dublin for favourable conditions. Amy, his wife of a few weeks, had come over in her own plane.

The weather then improved and at about 11.30 am on 18th August 1932 Jim Mollison took off on a history making flight. He had christened his plane *"Heart's Content"*. This had come about in a special way. The Automobile Association had prepared his map for the Atlantic crossing and drew his flight route with a bold purple line on a Great Circle course across Newfoundland. When it crossed the coast on the other side, it had passed over three closely situated Newfoundland villages named Heart's Content, Heart's Desire and Heart's Delight. Heart's Content was the village he hoped to fly over should he remain on course and accordingly he christened his plane after it.

The plane had no radio and with 162 gallons of fuel he had a maximum of about thirty three hours flying time. Mollison's full intention was to make a round-trip to and from New York. The plane was only a 102 horse-power machine. While daylight lasted he flew low and many ships saw him as he passed overhead, the last about 800 miles (1,287 km) west of Ireland. He drifted somewhat south and also met a lot of fog as he approached the Newfoundland coast. With great difficulty he eventually picked up Harbour Grace. He flew over the aerodrome but decided to carry on. He crossed the sea again and picked up Halifax, Nova Scotia and was now in the air for almost thirty hours.

He next flew on over New Brunswick and his fuel gauge there registered a usage of 152 gallons, leaving around ten gallons left. He realised by then that he would not reach New York and picked out a

Jim Mollison at Roosevelt Field, New York 28th August 1932

Photo No. 75-7785 *Smithsonian Institution*

safe landing place at Pennfield Ridge, New Brunswick where he put his little Puss Moth down in a meadow. Jim Mollison had become the first man to make a solo flight across the North Atlantic from East to West. He had made history. Word was flashed to his wife Amy that he had arrived safely.

Amy was delighted with his success and then it was her turn for fame when in November of that year 1932 she flew from England to Cape Town, beating her husband's time in the process by seven hours. The press headlines again had been captured. When she returned to England it was to a great welcome. For about six months Mollison and Amy lived in London where they attracted enormous attention everywhere they went. He was later to write "To say we lived in perfect accord on all occasions would be exaggeration."

It was time to fly again. For some time Jim Mollison had been seeking pastures new and in February 1933 he decided to become the first to fly from England to South America and also to become the first to fly across the South Atlantic. A very eminent astrologer in the Sunday Express had forecast the death of a prominent aviator around the time Mollison was to set off on his journey. The aeroplane he was using was the record breaker *"Heart's Content"*.

After take off, he flew to Spain where he stopped at Barcelona to refuel. He then continued southwards, passing over Gibraltar. Once he had passed Dakar he found that the Trade Winds were in his favour, blowing from Africa in the direction of South America. The heat now became intense as Mollison flew on over the South Atlantic. His course put him over St Paul's Rocks midway across the ocean but they were nowhere to be found. He was worried as he knew he must have drifted off course. On he flew and to his great relief he came upon the Placid Areas approximately 500 miles (804 km) off the South American coastline. This was a huge expanse of flat calm greenish blue water surrounded by rough Atlantic waves.

Mollison was now happy and flew on to land at Port Natal. He had flown across the South Atlantic from the shores of Africa in 17 1/2 hours. There he stayed for three weeks and then flew on southwards to Rio de Janeiro and on to Montevideo, the capital of Uruguay. He achieved the record that he had set out to make and the astrologer's forecast had not come to pass.

In July of 1933 the husband and wife aviators flew together from England to New York in an aeroplane they named *"Seafarer"*. The plane was a D.H. Dragon. Crossing the Atlantic they drifted off course and picked up their first landfall, the coast line of Newfoundland, at the

Straits of Belle Isle. They then set course for New York. It was hoped to be a record flight but twenty minutes flying time from New York they ran out of fuel. Mollison was forced to land as they were approaching Bridgeport Aerodrome. The plane landed on soft grass and turned two somersaults. Mollison was badly injured having undone his safety harness to give himself more room during the landing. Both he and Amy ended up in hospital. She was not badly injured but her husband had to get 107 stitches as a result of being thrown through the glass windscreen.

Despite their injuries they agreed to go to New York where a huge reception awaited them. They were taken there by ambulance plane and treated to a ticker tape reception four days later with a "Broadway Parade" of thousands. Because of crashing so close to New York in their record-breaking attempt, the New York Stock Exchange had subscribed sufficient money for the Mollisons to buy a new plane. However, Lord Wakefield again offered to replace the plane and they reluctantly refused the generous offer of the Americans.

In 1934, an Australian millionaire offered a prize of £10,000 to the winner of an air race from England to Australia. When it was announced , Jim Mollison was in London while his wife Amy was in Florida. De Havilland had offered to build three racing planes for those interested at a price of £5,000 each. The planes were called DH Comets.

Mollison was eager to compete again and, with a prize of £10,000 to be won, put a deposit on one of the planes. He cabled his wife telling her this and asked her would she come fifty fifty with him. She agreed and returned to London where they planned meticulously for the Melbourne Air Race. They called the plane *"Black Magic"*. It was capable of a speed of 230 mph. The had time to have only three short flights in the new plane before the actual race which started in October from Mildenhall. The draw put them first away and they flew to Baghdad. They were first to reach India in a record time of 22 hours. However, near Allahabad the plane developed engine trouble and they were forced to land. Three pistons had burned out. It was a bitter disappointment and one that didn't add harmony to the husband and wife team that appeared to be seldom on an even keel.

They had to withdraw from the race and it was three weeks before they could leave India to return home. They flew back across the deserts and eventually landed in Athens. From there Amy Mollison flew home by airliner while her husband flew "Black Magic" alone to England. He eventually sold the plane to the Portuguese Government.

In October 1935 when Jim Mollison returned from Singapore, he was only home a few days when his wife Amy set out on a record breaking flight to Cape Town and back. She returned in a blaze of glory to Croydon. He was there to meet her but it was to be their last public appearance under such circumstances. Their marriage by now was almost over.

In October 1936 Mollison had another record attempt in his sights. Leaving Amy behind in London he travelled to New York to fly to London and then on to Cape Town. His plane was an American Bellance Flash which he had never flown before and which was classed as unairworthy by the British authorities when the famous Irish aviator Colonel Fitzmaurice entered it on behalf of the Irish Sweepstakes for the Melbourne Cup in 1934.

Jim Mollison with his personal life in turmoil was not the greatest candidate to attempt such a crossing. News of the impending break up of his marriage had reached America before him and headlines such as "Flying Mollisons to split" greeted him. To try and reduce all the speculation, the stories and innuendo, Amy Mollison gave a press interview in the UK in which she said "I have decided it best, for personal reasons, that I and my husband should part. I want to live entirely separately from him. I want from now on to be known by my maiden name of Amy Johnson."

Despite all this, Mollison had the very serious matter of preparation for his flight to consider. He was extremely worried about his plane which he had christened *"The Dorothy"*. It was a difficult plane to fly and it had cost him £5600. Before his attempted crossing he had only flown it twice. The second flight was when he had brought the plane from Wilmington to New York for his North Atlantic flight. There was great interest in the intended attempt on both sides of the Atlantic, the marriage break up adding greater interest.

When he brought the plane to New York he had to wait about for a reasonable weather report. There was some hope that he might be able to get away at dawn on the third day of his stay there and instead of a good night's sleep beforehand he had a night "on the town". He was eventually contacted about four am that the weather was favourable and he set out straight away, still in his evening clothes. A friend gave him a raincoat to cover his dinner jacket. Press and camera men were all there to see him off. One observant reporter flashed a message to London "Mollison flying the Atlantic in evening dress."

He left New York at dawn and set course over the sea to Nova Scotia and six hours later landed at Harbour Grace, Newfoundland. It was bitterly cold when he arrived. He slept badly and the weather reports for a North Atlantic crossing were intimidating. Some were even conflicting but he decided to go, his reputation he felt was at stake. As he took "*The Dorothy*" into the sky the icy waters looked menacing below him. For a while he was flying through black clouds but then all around him became clear and he had an uneventful crossing, eventually landing at Croydon in a record time of 13 1/2 hours. There was one face missing from the welcoming crowd. It was Amy's and Mollison knew full well then that all was over between them. The flight itself had been a tremendous success and also a very lucrative one. He and Amy met a week later, there was no hope of reconciliation and they mutually agreed on separation and later divorce.

It was a sad ending to a romantic relationship between two great aviators. The high profile that Amy Johnson had as a woman aviator has often relegated Jim Mollison to one who had been just her husband. Yet, nothing could be further from the truth. He was an outstanding flyer. During their time together he not only achieved fame as being the first to make an East-West solo crossing of the North Atlantic, he was also the first man to fly solo East-West across the South Atlantic and was the first to cross both the North and South Atlantic. He died in 1959.

BALLOONS AND BALLOONISTS

Ever since man observed the birds in flight, he hoped that one day he would emulate them. Countless centuries however went by before the matter was given any serious consideration and then only in theory. The Greek mathematician Archimedes gave the subject some thought around 250 B.C. Roger Bacon, the English Franciscan monk in 1250 AD made a considerable contribution to the idea of man being able to leave the ground in some form of air machine. He wrote a book called *Secrets of Art and Nature* in which he referred to "Engines for flying, a man sitting in the midst thereof by only turning about an Instrument which moves artificial wings made to beat the air much after the fashion of a bird."

Although aviation did not evolve in that way, it was six and a half centuries later before a heavier-than-air machine was first flown. In the intervening period there were many contraptions used to get man into the skies. One of those was the kite, the origin of which goes back long before the time of Christ. The Chinese are credited with flying kites about 300 B.C. The use of a man-lifting kite did not come to notice until the 14th century when Marco Polo recorded seeing them being used for that purpose in China. Lawrence Hargrave of New South Wales, Australia (1850-1915) by the use of box kites undertook a very detailed study of aeronautics and so contributed greatly to the future development of the aeroplane.

It was the balloon more than anything else however that gave man that freedom to take to the skies. The distinction of being the originators of the hot- air balloon has very often been credited to the Montgolfier Brothers of France in the 1780s but as early as 1709 it was demonstrated by a Brazilian priest Fr. Bartolomeu Gusmão. On 8th August of that year in the Ambassador's drawing room in Lisbon he showed King John V, Queen Maria Ann and other notables including Cardinal Conti who later became Pope Innocent III how a model air balloon could lift off if the air was heated from beneath it. The experiment was very successful and all those present were extremely impressed. Father Gusmão's demonstration, although only with a model , clearly showed that if such an experiment could be carried out successfully with a much larger balloon out of doors , then there was no reason why eventually man could not be lifted from the ground and transported elsewhere.

On 25th April 1783 at Amonay near Lyons in France, the first of such balloons became airborne to a height of 1000 ft. (305 m) and credit for this went to the Montgolfier brothers Joseph and Etienne. Further developments went ahead rapidly and the brothers gave a display of their invention to King Louis XVI on 19th September 1783. At this event, the Montgolfiers fitted a basket underneath the balloon and in it they placed a duck, a sheep and a cock. The balloon reached a height of about 1800 ft. (550m) much to the wonderment of the King and his wife Marie Antoinette.

Ballooning was now on its way and it was felt that soon man would be able to take to the skies. About a month later, a Montgolfier designed hot-air balloon with a diameter of 49 ft (15 m) and fitted with a cradle capable of holding two men was used in a controlled flight to carry aloft a young Frenchman named Francois Pilâtre de Rozier. The balloon was restricted in the height it could go by a rope attachment to the ground. It lifted to 85 ft (26m) and stayed aloft for 4 1/2 minutes with de Rozier on board. On 21st November 1783 de Rozier and the Marquis d'Arlandes became the first in the world to travel by balloon in free flight. They flew for approximately 25 minutes and in that time had travelled over 5 miles (8 Km).

France was the early home of ballooning and the next major advance in this type of aviation was the use of hydrogen instead of air. Hydrogen is approximately 15 times lighter than air at standard pressure and temperature. This gas was initially called Phlogiston. Credit for its discovery in 1766 goes to the British scientist Henry Cavendish. The hydrogen balloon quickly replaced the hot-air balloon and within weeks of that epic flight of 21st November, Professor Jacques Charles and one of the Roberts brothers were the first to be carried in free flight by a hydrogen balloon. This took place on 1st December 1783 from Les Tuileries Gardens in Paris, before an estimated crowd of 400,000. The balloon landed 27 miles (43 Km) away near the town of Nesles.

Ballooning developed rapidly and soon ascents were being made in many countries in Europe and in both England and Ireland. On 25th November 1783 the first balloon demonstration was given in London by an Italian Count Francesco Zambeccari. It was a hydrogen balloon of small dimensions, 10 ft. (3.05m) in diameter. The first unmanned hot-air balloon flown in Ireland was in February 1784 by a man named Riddick who gave the demonstration from the Rotunda Gardens in Dublin.

On 4th June a French lady named Madame Thible became the first woman in the world to be carried in free flight in a Montgolfier-designed hot-air balloon called *Le Gustav*. She flew from Lyons. Also on board were Monsieur Fleurant de Rozier and Joseph Montgolfier.

On 15th September 1784 an Italian, Vicenzo Lunardi of Lucca ,who was attached to the Italian Embassy in London made the first flight in England in a hydrogen balloon.

No sooner had man discovered the ability to leave the ground and float aloft than he saw the enormous potential that ballooning offered. New frontiers could be conquered, seas could be crossed and the balloon in time of war could have its uses. The English Channel was quickly seen as a challenge and on 7th January 1785 the first aerial crossing was made by Frenchman Jean-Pierre Blanchard and American Dr. John Jeffries. They took off from Dover and 2 1/2 hours later landed at Foret de Felmores. On the way they began to experience difficulty in maintaining height and to reduce weight they were obliged to discard most of their clothing. It was inevitable that tragedy would strike among the ballooning fraternity and sadly on 15th June 1785 Francois Pilâtre de Rozier and Jules Romain became the first victims. They had been trying to cross the English Channel from Boulogne in a composite hot-air/hydrogen balloon when they crashed. It was felt that this was due to the venting of hydrogen which caused a spark.

Although the French dominated the balloon scene for some time and produced some notable balloonists, other countries also had their ballooning heroes. Ireland had Richard Crosbie. He was born in 1755, the son of wealthy parents and entered Trinity College, Dublin in 1773. Ten years later when the first free flight man-carrying balloon took to the skies, its success aroused great interest in the very technically-minded Crosbie who began experimenting. By August of the following year he had devised what was then referred to as an Aeronautic Chariot which he demonstrated. Crosbie was described by Sir Jonah Barrington, an eminent lawyer, as having a brilliant mind and who was "beyond comparison, the most ingenious mechanic I ever knew. He had a smattering of all sciences, an interest not shared by his father who did everything in his power to curb the young man's enthusiasm."

Richard Crosbie's ambition to become the first man to cross the Irish Sea in a balloon evaded him. His first attempt was on 19th January 1785. The take off point was Ranelagh Gardens in Dublin where it was estimated that over thirty thousand people saw him lift off in a hydrogen balloon at around 2.45 pm. However, due to the approach of darkness at that time of year, he decided not to fly out over the sea and so came down again at Clontarf. He made a second attempt at the crossing on 10th May but the balloon barely lifted off due the excessive weight of paraphernalia he was carrying in the basket. So as not to disappoint the crowd he got a lightweight student from Trinity College

Mr. Crosbie's Ascent from Leinster Lawn.

National Library of Ireland.

named Richard McGwire to go in his place. It was anticipated that he would never make it and he did in fact come down about nine miles out to sea where he was picked up by a passing fishing boat. Richard Crosbie made his last attempt to cross the Irish Sea on 19th July but was forced to come down half way across and was picked up by a barge.

James Sadler was the first English aeronaut. He flew in a Montgolfier type home-made hot-air balloon near Oxford on 4th October 1784. He was injured the following year in a ballooning accident while flying from Manchester and was forced to abandon further flights. However, in October 1812 at the age of 61 he set out on a balloon flight across the Irish Sea with the intention of becoming the first to do so. After two hours he discovered that he was close to the south coast of the Isle of Man and hoped to reach Liverpool. Unfortunately this wasn't to be and he was forced to come down in the sea where he was picked up. On 22nd July 1817 his son Windham Sadler made the first aerial crossing of the Irish Sea when he flew in a balloon from Portobello Barracks in Dublin to Holyhead in Wales.

Despite the intense activity of ballooning in Europe, it wasn't until 9th January 1793 that the first untethered balloon flight was made in America by the French balloonist Jean-Pierre Blanchard. It will be recalled that eight years previously Blanchard made the first balloon crossing of the English Channel. In America he took off in a hydrogen balloon from Philadelphia and 44 minutes later landed about 15 miles away in Gloucester County, New Jersey. An interesting point is that it was witnessed not alone by President George Washington but also by four future presidents of the US - Adams, Jefferson, Madison and Monroe.

While ballooning advanced quickly in America following Blanchard's flight, its military feasibility was not seriously considered for some time. In Europe however, the military potential of the balloon was quickly seized upon and France became the first country to use a man-carrying balloon for military purposes. In June 1794 the French Republican Army deployed a tethered military balloon named *Entreprenant* at Maubeuge, Belgium during the battle of Fleures against the Austrians. Capt. Coutelle made an ascent in the balloon to survey the enemy and it was claimed that the observations made were of considerable advantage in the defeat of the Austrians. Balloons were used to some extent for surveillance in the American Civil War and a Balloon Corps was formed. It only lasted a short while before it was disbanded and the operation of the balloon for military purposes was not again taken seriously in America until the early 1890s. Military use

James Sadler Esq.
First English Aironaut.

Engraved for the Dublin Magazine.

of balloons in England did not come about until 1878 when the army had one constructed. They were later used extensively by the British for observation purposes in the First World War.

The future potential of the balloon for non military purposes, as for example long distance travel, gained considerable impetus due to flights such as that on the 7th-8th November 1836 when a hydrogen balloon named *The Royal Vauxhall Balloon* with a crew of three lifted off from Vauxhall Gardens in London and travelled a distance of 486 miles (772 Km), landing at Werlberg in the Duchy of Nassau. In July 1859, a distance of 1120 miles (1800 Km) was covered in America by balloon in a flight from St. Louis to Henderson, New York. The one great disadvantage of a balloon is that one is forced to travel only in the direction of the wind.

Eventually the airship, a lighter than air craft was developed where steering and propulsion overcame the vagaries of free flight and the balloon faded out until there was somewhat of a revival in the 1960s but not for any commercial or military purpose.

In the meantime the balloon continued to be used in scientific research such as weather observations etc. It was also used in sport and there were many ballooning events held, such as the Gordon Bennett Races hosted in some European countries and also in the US where competitors competed to fly the longest distance from a starting point. These races spanned a period between 1906 and 1938, excluding the First World War years.

In recent decades, interest in ballooning not alone as a sport but also as a method of breaking aerial records has developed considerably. Although the first attempt to cross the Atlantic in a hydrogen balloon was made in 1873, the crossing by gas balloon wasn't successfully undertaken for another 105 years when between the 12th and 17th August 1978, Maxie L. Anderson, Larry M. Newman and Ben L. Abruzzo crossed the Atlantic in a balloon named *Double Eagle II*. It was a journey of 3107 miles (5001 Km)

In more recent times 2nd-3rd July 1987, the first Atlantic crossing by a hot-air balloon *Virgin Atlantic Flyer* was made by Richard Branson and Per Lindstrand from Sugar Loaf, Maine, USA to Limavady in Northern Ireland. The flight achieved the distinction of being the fastest manned hot-air balloon flight ever.

Ballooning as an early form of aviation laid the foundation for future developments such as the airships but these lighter-than-air giant craft found little or no place in commercial aviation following the disasters of the R101 and the Hindenburg and so too faded into aviation history.

INDEX

Adelt, Leonhard. 134, 136
Aerial Register & Gazette 117
Aeronautical Journal, London 70
Aeroplane, The 37
Alcock & Brown 10, 84
Anderson, Maxie L 156
America - Flying Boat 84, 85
Army Vet. Corps. 116
Arnold, Gen. 109
Atherstone, Lt. Cdr. 38
Atlas Corporation 111
Autogiro 126

Bacon Roger 150
Baldonnel 14, 21, 26
Ballcher, Bernt 48
Barton, George 14
Beaverbrook, Lord
Behr, Kennett 21
Bell X-1 111
Bellance Flash 148
Bellinger, Lt. Cdr. P.N.L. 88, 92
Bendix Trophy 108, 110
Bennett, Floyd 18, 21, 22
Berkshire Aviation Co. 118
Bird, Commander 45
Bishop, Billy 104
Black Magic 147
Black, Tom Campbell 126
Blackburn, Robt. 155
Blanchard, J.P. 152, 154
Blanco, Josephine 62
Booth, Sqdn. Ldr. 36
Brackner, Sir S. 36-39, 120, 124
Branson, Richard 156
Breeze, James 89, 90
Bremen 9-16, 18
Brimstead, Col. 141
Bureau of Air Commerce 24, 28
Burks, Bill G. 62
Burry Point, Wales 46
Byrd, Lt. R.E. 85

C-5 Airship 88, 90

Caldwell, Capt. K.L. 103
Callaghan, Lt. 104
Canavan, Leo 14
Caribou - Flying Boat 127
Cavendish, Henry 151
Century Magazine 78
Chamberlain, Clarence 10
Chanute, Octave 69-72, 74-76, 78, 82
Charles, Prof. Jacques 151
Chiang Kai-Shek, Madame 110
Chicago Daily News 80
Chicago, USS 90
Churchill, Winston 94
Coachella Ranch 108
Cobham Aviation Ltd. 124
Cobham, Sir Alan 116-127
Cobham-Blackburn Airlines 125
Cochran, Jacqueline 60, 106-115
Colliers Magazine 136
Colmore, Wing Cdr. 31, 33, 36-39
Colorado, USS 59
Columbia, USS 91
Corrigan, Douglas 21-28
Cosgrave Pres. 15
Cowley, Countess 143
Cowper Swift 126
Cripps, Sir Stafford 40
Crosbie, Richard 152, 153
Cruise, Vice-Adml. E.A. 62
Cudahy, John 27, 28
Curling, Capt. Richbel 67
Curtiss Robin 21, 22
Curtiss, Glen 84, 85

d' Arlandes, Marquis 151
Daily Mail 84. 86
Daniels, US Navy Sec. 86
DC-3 128
De Havilland Aircraft Co. 119, 120, 122, 123, 147
de Havilland, Geoffrey 119, 123, 124
de Rosier, F.P. 151, 152
Deighton, Len 137
Denison, Harold T. 45

DH Puss Moth 143, 144
DH2 99
DH.9C 119
DH.5O 122,123, 124
DH.60 119
DH. 61 124, 125
DH. Dragon 146
Dorothy ,The
Double Eagle II - Balloon 156
Douglas Aircraft Co 128.
Dowding, Air Vice- Marshall 34, 36
Doyle, Sir Arthur C. Doyle 37

Earhart, Amelia 44-63
Earhart, Edward 44
Eckner, Dr. Hugo 129, 130, 136, 137
Eyles, Mr. & Mrs. 96, 97, 100, 103

FAA 29
Fayal Island 91
Ferber, Capt. 74, 80
Fitzgerald, Desmond 15
Fitzmaurice, J.C 9-10, 12-20, 148
Flanagan, Sister 103, 105
Flight Refuelling Ltd. 126
Flyer II 83
Fonck, Rene 95
Ford, Henry Museum 18
Friendship , The 46, 50

Garrett, Eileen 37
Giffard, Henry 30
Gladstone, Capt. Tony 125
Gordon, Lou 46
Gower, Pauline 109
Graf Zeppelin 128-130, 137, 139
Greenly Island 116
Grieve, Mac Kenzie 92
Guest, Mrs. Frederick 45, 46
Gusmäo, Fr. B. 150

Handley Page W.10 126
Hargrave, Lawrence 150
Hawker, Harry 92
Heart's Content 146

Heath, Lady Mary 47, 64-67
Heath, Sir James 66
Henson, Everett Jnr. 62
Higgins, Sir John 33, 34
Hinchliffe, Capt. W. 10, 36
Hinchliffe, Emily 36
Hindenburg 128-139
Hinton, Lt. Walter 90
Hoare, Sir Samuel 124
Hogan, Major Gen. 15
Homes, Fred & Jack 117
Howland Island 54, 56, 59, 60
Hull, Cordell 28
Hünefeld, Gunther Von 9-17, 20

Imperial Airways 125
Inglis, Prof. 40
Irish Air Corps 9, 12
rwin, Flt. Lt. 29, 31-33, 37-40, 42
Irwin, Mrs. 40
Itasca USS 56, 58-60

Jackson, Justice Robt. 110
Jeffries, Dr. J. 152
Johnson, Amy 140, 141, 143, 144, 146-148
Johnson Col. J. M. 28
Johnson, Sqdn. Leader 37, 38
Junkers 10, 11

Kinner Canary 45
Kinner, Wm. 45
Kitty Hawk 70, 72, 76, 80
Klose, Herr 11
Köhl, Captain Herman 9-15, 17, 18, 20

Langley, Samuel P. 75
Lau, Helmet 133
Leech, H.J. 39
Lehmann, Capt. 131, 134, 138
Levine, Chas. 10
Lexington, USS 59, 60
Lilienthal, Otto 69, 74
Lindbergh, Chas. 10, 84
Lindstrad, Per 156

Lloyd, Gladys 118
Lockheed F-104 114
Lockheed Lodestar 114
Lockheed Vega 48, 50-53
Lockheed 10E-Electra 52, 54-58, 63
Loose, Herr Fritz 10
Love, Nancy 109
Lufthansa 11
Lunardi, Vicenzo 152
Lynn, Elliott 64
LZ1 128
LZ129 129
LZ130 130. 131, 139

Mackay, Elsie 10, 37
Maher, Johnny 14
Manning, Capt. H. 52, 54
Mannock, Major E. "Mick" 95-105
Mannock, Mrs. Julia 95
Mantz, Paul 50, 52, 54, 55
Marshall, Ted 107
Mather, Mgt. G. 133
McArthur, Gen. 110
McCartney, Det. G. 138
McCudden, Lt. James 98, 99, 103
McDonald Ramsay 30
McIntosh, Capt. 10
McWade, Inspector 33, 42
Meagher, Capt. 32
Melbourne Air Race 148
Miasaouri
Miss Colombia 83
Mollison, Jim 140-149
Montgolfier Bros. 150, 151
Mooney, Ml. M. 138
Moore-Brabazon, Col. 40
Morgan, Wm. 92
Morrison, Herbert 132-134

National Aviation Day 126
NC-1 85, 88, 90, 92
NC-2 86
NC-3 88, 90-92
NC-4 84, 88-91, 93, 94
Newman, Larry M. 156

Nicols, Ruth 47
Noonan, Fred 44, 52, 54-56, 58-61
Northcliffe, Lord 84, 86

O'Kelly, Pres. Sean T. 28
Oakland Tribune 61
Odlum, Floyd 106, 108, 111, 114
Olds, Gen. 109
Ontario USS 56, 58

Pan American Airways 52, 55
Peacock, Sid 14
Pierce, Dr. George 64
Pierce, Jackie 64
Pierce, Sophie 64-67
Polo, Marco 150
Porte, John Cyril 84
Portmarnock Strand 144
Pruss, Capt. 133, 138
Putnam, George P. 45, 47, 48, 54, 59, 60

R-33 30
R-34 30, 94, 128
R100 31-33
R101 30-43, 126, 128, 137
Railey, Capt. Hilton 45-47
Raynham, Frederick 92
Read, Lt. Col. A.C 84, 87-93
Rhoads, Eugene 90
Richardson, Cdr. H. C. 85, 88
Richmond, Col. 31
Richtofen, Baron Von 95
Rodd, Ensign Herbert 90
Rogers, Capt. Kelly 127
Romain, Jules 152
Roosevelt, Mrs. E. 50. 59, 109
Roosevelt, Pres. 50, 59
Roper, Daniel C. 136
Royal Air Force 99, 123, 124, 140
Royal Army Medical Corps 96, 97
Royal Flying Corps 9, 36, 98, 117

Sabre F-86 112
Sadler, James 154, 155

Sadler, William 154
Santos-Dumont. D.A. 71, 83
Schiller, Capt. Hans Von 137
Schute, Nevil 32
Scott, C.W.A. 142
Scott, Major 31, 37, 38
Seafarer 146
SE5A Fighter Plane 102
Short Bros. 81, 83
Siddely, Sir John 122
Sikorsky 108
Simon, Sir John 40-43
Smith, Chas. Kingsford 141, 142
Southern Cross 141
Spalding, Capt. 44
Spindler, Lt. 12
Spirit of St. Louis 21
St. Elmos Fire 137
Stone, Lt. Elmer 90
Stults, Bill 46
Swan USS 56

Taylor, Charlie 75
Thibble, Madame 151
Thomas, Meredith 99
Thompson, Lord 32, 33, 38, 39, 42, 120

Tompkins, Eric 98
Towers, Lt. John 84, 85, 90-93
Trenchard, Air Chief Gen. 102
Tweed, Major 40

Vandenberg, Gen. 112
Vickers Aircraft Co. 31
Virgin Atlantic Flyer 156

Wakefield, Sir Charles 122, 144, 147
WASPS 109, 110
Wellesley, Lady Diana 143
Wellington, Duke 143
Westcott, Cyril 142
Westervelt, Cdr. G.C. 85
Wilson, Pres. W. 94
Wright *"Flyer"* 68, 76, 83
Wright, Orville 68-69, 74-77, 78, 80, 83
Wright, Wilbur 68-78, 80, 83

Yeager, Chas. "Chuck" 111, 112, 114

Zambeccari, Count F. 151
Zeppelin Aircraft Corp. 129
Zeppelin, Count F. Von 24, 128